Modeling Engineering Systems

PC-Based Techniques and Design Tools

by Jack W. Lewis

publications inc.

Solana Beach, California

This is a volume in the Engineering Mentor™ Series

Printed in the United States of America
10 9 8 7 6 5 4 3 2

Cover design: Brian McMurdo, Ventana Studio, Valley Center, CA
Developmental editing: Carol Lewis, HighText Publications
Interior design: Sara Patton, Wailuku, HI

Library of Congress Cataloging-in-Publication Data

Lewis, Jack W., 1937–
    Modeling engineering systems : PC-based techniques and design tools / by Jack W. Lewis
       p.   cm. – (Engineering mentor series)
    Includes bibliographical references and index.
    ISBN 1-878707-08-6 : $19.95
    1. Mathematical models–Data processing. 2. Systems engineering–Mathematical models–Data processing. 3. Computer-aided engineering.
I. Title. II. Series.
TA342.L48 1994
620'.001'1–dc20
                                         93-40319
                                            CIP

publications inc.

P.O. Box 1489 ■ Solana Beach, CA 92075
HighText is a trademark of HighText Publications, Inc.

This book is dedicated to Dr. Herbert H. Richardson, a professor of mechanical engineering at MIT during my graduate school years of 1964–66. H^2R, as he was fondly called, probably taught me more about engineering than all the other teachers and professors I had, combined. He was an outstanding teacher who seemed to have a hundred different ways of explaining a complex subject in a simple and humorous manner. I owe much of my success in engineering to this man.

~

Contents

Preface

If you're like me, when you pick up a book you like to know something about the person you're trying to learn from. Quite simply, I'm a practicing engineer who believes anyone can learn engineering, as long as it is explained by someone who is willing to take the time to present the subject matter in simple enough terms. That's what this book is about.

I graduated from the US Coast Guard Academy in 1960 with a B.S. in Science. I did well there—at least I got good grades and was near the top of my class. After four years at sea, I was selected to go to graduate school at MIT. What a humbling experience that school was for me! During the first semester, I realized that I wasn't nearly as clever as I thought. In fact, I quickly became convinced that I knew nothing about math and science, and that I was going to flunk out!

However, as I gave up the notion that I knew anything, the situation gradually began to change. At MIT I encountered, for the first time it seemed, professors who were not trying to impress me with their knowledge and intelligence.

They sincerely wanted me to learn. When I wasn't understanding something, they took it as a failing on their part, not mine. It seemed that most professors had 50 different ways to teach a fundamental principle.

At MIT, I learned engineering in a way I will never forget, but most importantly, I learned that engineering is about *solving practical physical problems by creating mathematical models that can be manipulated.* These models allow you to learn a great deal about the physical problem that you're trying to solve. Engineering is not just mathematics—rather, mathematics is simply a tool used in engineering.

Through my coursework at MIT in the area of automatic control systems, I learned that all engineering systems look alike mathematically—a concept that is at the heart of this book. Incidentally, I did manage to graduate from MIT with a master's degree in mechanical engineering and a naval engineer degree. I started my own consulting engineering business in 1970. I'm still active in the consulting arena, and still love engineering.

Chapter

1

Read Me

1.1 The Engineering Design Process in a Nutshell

I've discovered during the writing of this book that it's important to know who your readers are. As a reader, it's also important for you to know who this book is aimed at. I would like to think that anyone with a strong desire to learn the fundamentals of engineering can do so in the pages of this book. I do believe that, actually, but it will help a lot if you already have a strong foundation in mathematics—at least one or two semesters of calculus.

This book is not a textbook per se. It is primarily intended for those just beginning their engineering careers, or for practicing engineers who are changing fields or who need a brush-up. I also believe that senior technicians who want to press ahead into more advanced engineering design work can also benefit from it. For those whose math is a little rusty, Appendix A provides a "quick and dirty" review of both differential and integral calculus. Other mathematical topics, such as complex number theory, are introduced and explained in the text as they are needed.

Before we get into the nuts and bolts of modeling engineering systems, bear with me for a few more moments while I wax philosophical about the engineering process. My definition of engineering is the application of physics and other branches of science to the creation of products and services that make the world a (hopefully) better place. Your "success" in engineering will likely be closely related to how well you can create products and services that your organization's customers need and want. Unfortunately, I have found that creating new products and services is a lot more difficult than analyzing or criticizing those that already exist. There are far more critics in the world than creators!

Both a creative and a critical person are inside each one of us. To be good in engineering, you have to be able to "turn off" your critical side long enough to allow yourself to create. Once you have created, then you can turn the critical side back on to analyze and pick apart your creation. Watch out for all those other critics in the world. If you constantly listen to them, you will never succeed in engineering (or anything else, for that matter). Learn to encourage yourself. The praise of peers will follow.

This book is about *modeling* and *analyzing* engineering systems. Modeling is the creative side of engineering, and analyzing is the critical side. I use the term "engineering system" in this book to refer to a product or device that may contain mechanical, electrical, fluid, and/or thermal components. An engineering system can therefore be interdisciplinary, and require a designer to have knowledge of many engineering fields.

Creating an engineering design does not have to be a mysterious art. The more you learn about what is available in the way of real-world basic components and services, the more creative you will become. This book contains all of the fundamentals needed to develop mathematical models of engineering systems and to analyze these models. But you must make it a habit to collect and carefully study product and service catalogs of basic components so you know what materials are readily available. Don't be afraid to ask "dumb" questions. Product representatives as well as more experienced engineers and technicians can be the source of a great deal of knowledge. It sounds like a cliche, but asking questions is really the easiest way to learn. Then, when you are trying to create a new design, you will have a lot of information in your head that will (almost subconsciously) help you generate ideas.

Figure 1.1 shows the basic process undergone in developing a new engineering design. The design is created first in your mind, when you become aware of a want or need. You then make a list of specifications that you envision would satisfy that need. Then you will create (in your mind, at first) a product or service that meets these specifications.

In order to "give birth" to this new design idea, you have to get it out of your mind and into reality. Take care at this point. Don't let your (or anyone else's) critical side take over too soon, or your ideas will never be allowed to take root.

You can transfer your "mind model" into a "symbolic" or even a "physical" model. It is usually best to work with symbolic models at first, because they are typically less expensive than physical models (but not always). A symbolic model might be a circuit diagram from which you can derive a number of mathematical expressions describing the behavior of your mind model. You can then solve these mathematical expressions for the answers you need.

Once you have the first solution to your mathematical expressions, then—and only then—allow your critical side to tear apart your design. Don't be discouraged if the first attempt at meeting the specifications seems silly. Use the

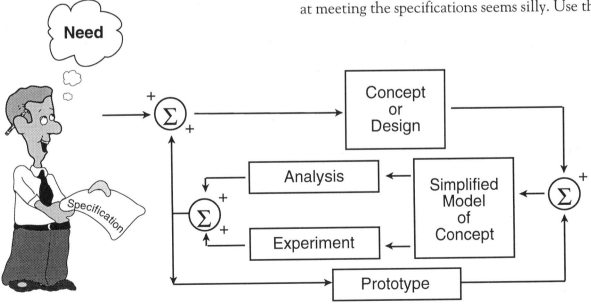

Figure 1.1. The engineering design process.

results of your analysis to see how well your product met the specifications, and alter your specs and modify the design as needed. Then create another model to analyze. Continue to iterate in this manner until you have a product that satisfies the need.

At this point in your design you have to ask yourself, "Am I confident enough in my analytical results to proceed with the construction of a full-scale prototype?" If the answer is no, then consider building and testing a physical model of your product. Physical model experiments can produce empirical solutions to problems that are difficult to solve mathematically or for which too many untested assumptions had to be made in order to derive the analytical model and its solution. Building and testing physical models are usually much less expensive than building and testing a full-scale prototype.

Results of physical model-testing experiments could lead to a revision of your specifications, a new mathematical model, or even another series of physical model experiments. Eventually you will get to the point where you are confident enough with your design to proceed to the construction of a prototype. Seriously consider at this point conducting experiments with this prototype to obtain data to verify your analytical and physical model results. If the prototype does not meet your original specifications, you may have to iterate once again through the design cycle before proceeding further.

This is the engineering design process in a nutshell. It may sound long, complicated, and expensive. It can be, when major, complex engi-neering systems are involved. But no matter how big or small the system, *never* short circuit this engineering design process! Engineers and technicians are not artists who create their masterpieces on canvas with a brush, or out of marble with a hammer and chisel, with little input from others. Instead, they are people who design, redesign, and then let others check their designs. They convert their designs into engineering drawings and then they allow skilled craftsmen to provide feedback on how to improve the design to reduce production time and costs. I have seen and worked with too many "seat-of-the-pants" engineers who have the engineering design process backwards. They first build prototypes and then they design, often with disastrous results. If you do this, you may be considered a good "artist," but you will never be a good engineer.

1.2 An Engineer's Tool Box

Any skilled craftsman knows that a good set of tools and the knowledge to use them is of fundamental importance in getting a job done properly and safely. An engineer also has "tools." Like the craftsman, some of these tools are physical in nature, but for the most part an engineer's tools consist of mental skills developed through study of mathematics and science.

This book will explain the fundamental mathematical and scientific tools needed to succeed in engineering. It will also show how to use them in practical applications. The tools are mostly mathematical in nature, because mathematics is at the core of engineering. *Please bear with me through the mathematics.* I have attempted to explain each step and to make it as simple and understandable as possible. But significant effort on your part is required.

I have assumed in the book that you have a fairly good grounding in algebra and trigonometry, and have had courses in both differential and integral calculus. If you find yourself getting "lost in the algebra," then brush up using a high school text. Calculus fundamentals are reviewed in Appendix A.

When reading a new engineering text, I usually keep a notebook handy. As I encounter new equations in the text, I jot them down and work through each one until I understand exactly what the author is doing. I highly recommend this method—although it makes for slow reading, it ensures that you understand the material covered.

When I started my engineering career, an engineer's physical tools consisted of a drafting board and supplies, a slide rule, engineering reference books, and textbooks. Many of the engineering reference books contained log and trig tables to help in mathematical calculations. There were no electronic calculators, and digital computers were kept in caged air-conditioned rooms where only the computer folks were allowed.

The electronic calculator and personal computer have completely replaced many of the physical tools that an engineer used to use. The hand calculator made the slide rule and many of the tables in engineering reference books obsolete. It's hard to imagine being without a good calculator.

The personal computer has revolutionized engineering. There are now so many powerful engineering and mathematical programs available for the PC and Macintosh that I think it is fair to say that the capabilities of these packages have surpassed the capabilities of the average user. I see many technicians and engineers using or attempting to use engineering and mathematical software packages who do not understand the fundamentals. For example, dynamic system simulation (at the heart of engineering) and DSP (digital signal processing) software packages are available now that simply astound me with their capabilities. To put these programs to good use, however, you must have the mathematical and engineering fundamentals. That is what this book is all about.

I have therefore assumed that you either have, or have access to, a personal computer. I also have assumed that you know how to use spreadsheets such as Excel, Lotus 1-2-3, or Quattro, and that you know how to use a higher level computer programming language like BASIC. (Spreadsheets are a surprisingly useful engineering tool. I use them frequently to develop very complex simulations.) If you have some of the latest simulation and DSP packages, that's great too. But, let me give you a word of caution. Having a computer and knowing how to run canned engineering programs doesn't make you an engineer any more than carrying around and knowing how to use a slide rule made you an engineer years ago. The latest and fastest computer and the latest version of a software package are the "trappings" and "images" of engineering—they are not engineering. The fundamental knowledge presented in this book is absolutely necessary to make the most of any engineering design software package on the market.

1.3 Analog Computers – An Anachronism?

When I first entered graduate school, I was introduced to an analog computer. What an incredibly powerful machine it was! It could solve differential equations much faster than a digital computer could at that time (and it still can). In fact, an analog computer solves dynamic system problems instantaneously, something no digital computer can do. So don't get the idea that an analog computer is "old fashioned" and that digital computers are the only computer you need in your engineering tool box.

In this book you will study operational calculus and block diagrams. These concepts are particularly useful in building mathematical models that can be solved on analog computers. Once an operational block diagram of the dynamics of an engineering system has been constructed, it is easy to put it on an analog computer and let it solve the problem. You can also build electrical analogs of physical systems and interface these to a computer without the need for expensive sensors. Then you can quickly develop digital control algorithms.

You may be wondering where you are going to find an "old" analog computer to do that. While it is true that you can't readily buy an analog computer these days, it's not because they are old and obsolete. It's simply because they were replaced by the operational amplifier—an integrated circuit that became so cheap and easy to use that anyone could build an analog computer for peanuts. Unfortunately, it seems that many educators have forgotten this, as I have run into young engineers, even electronics engineers, who have no idea how to use

operational amplifiers to build analog computers to solve differential equations.

A few years ago while I was studying the problem of controlling a water wavemaker (see Chapter 7), I built an electrical analog of the wave tank and associated hydraulic piston and valve so I could study a digital feedback controller. The software engineer working with me was amazed. How could a breadboard full of op amps behave like a water wavemaker?

You will learn in this book that the dynamics of all engineering systems—whether electrical, mechanical, thermal, or fluid—can be described by the same mathematical equations. That means mechanical, fluid, and thermal systems, which are difficult and expensive to construct and test, can be converted into electrical circuits, which are cheap and easy to test. This is the whole fundamental concept behind the "analog computer." The word "analog" really doesn't have anything to do with electronics, even though it is generally accepted these days that analog is anything that is not digital. The word analog is used to describe a system that behaves like another system yet has a different physical form.

Today you can go to Radio Shack or an electronics mail-order house and buy IC operational amplifiers, resistors, capacitors, and other components for pennies. All you need is a little skill, a power supply, a breadboard, and some wires, and you can build an analog computer that can solve mechanical, fluid, and thermal dynamic system problems as well as electrical circuit problems. Once you have such an analog model of your system, you can also very easily conduct experiments with it using a digital computer equipped with an analog-to-digital converter. Since your analog model will already be in electrical form, ready to interface to your digital computer, you won't have to go out and buy expensive sensors that convert mechanical, fluid, and thermal variables into electrical signals.

1.4 A Few Words About Units

Throughout this book I use the English system of engineering units, with the corresponding SI units in parenthesis. I do this because, in spite of efforts made to convert the United States and some other leading industrial countries to the SI system of units, I find that engineers stick to the units they "think" in. Let me explain with a story. In the early 1970s I made a concerted effort to get everyone in my firm to convert to the SI system. Then we got a job in Japan and I was happy that we had converted, since I knew the Japanese used the metric system and assumed they had made the easy conversion to the SI system. When I made my first presentation, much to my amazement I found that our Japanese clients (who were ship-builders) did not understand the units I was using and asked me to explain what they meant in terms of units they were familiar with. The units they used turned out to be a rather strange system that I can only call a "Japanese naval architectural metric system." Even today, the ship-building industry throughout the world uses some of the strangest sets of measurement units you will ever encounter, and I doubt they will ever change to the SI system.

Just how important are units in engineering? First of all, never lose sight of the fact that the laws of nature have no inherent system of units. Units are man-made. Indeed, if you run into an equation that cannot be made unitless (that is, dimensionless or without dimensions) by dividing through by some combination of variables, then the equation does not truly describe nature and is probably wrong or valid only over a very small range of the independent variables. Nondimensionalizing an equation is a good way to check your equations and an excellent way to present your results.

Of course, units are important in engineering. Many components you purchase will have weight, volume, or linear dimensions, and will consume or require power, produce a force or a torque, and so forth. These components may come from every part of the world and will make use of every conceivable system of units known to man. What do you do? You simply need to know how to convert from one set of units to another. If you can't think in the supplier's units, convert them to the ones you are familiar with or intend to use in your product. There are no right or wrong units to use. Just make sure your customers readily understand the units *you* use.

1.5 Overview of Book

Now—finally—I'll tell you a little bit about the content of this book and the way it's organized.

Chapter 2 is probably the most important chapter. You will learn that there are only three basic types of engineering building blocks, two that can store energy and one that dissipates energy. The concepts are deceptively simple, yet extremely powerful.

Chapter 3 begins the process of teaching you how to build math models of any engineering system. It covers all of the model-building techniques that have been developed over the years in many branches of engineering. You will discover the mighty first-order linear ordinary differential equation, the very backbone of engineering. From my experience, learning how to build math models is the hardest part of engineering, because it involves creating. I have spent a great deal of time on this chapter and tried to make it as easy as possible to grasp.

Chapter 4 introduces you to the analytical side of engineering. You'll learn how to solve the first-order linear differential equation math models you developed in Chapter 3. This is also a very important chapter. First-order linear differential equations can be used to describe many facets of engineering, from the flow of ground water through porous soils to the flow of electrons through electrical circuits. The so-called *time and frequency domain* solutions to these equations form the basis of all engineering analysis.

Chapter 5 is one of two chapters in the book intended to help you bridge the gap between theory and the solution of practical engineering problems. A problem is selected that involves modeling and analyzing a combined mechanical and electrical engineering system. The example helps emphasize that all engineering systems look alike mathematically. It shows how two first-order linear differential equations lead to a second-order linear differential equation, the subject of the next chapter.

Chapter 6 introduces you to the next important subject, second-order linear differential equations. This chapter addresses both the modeling and analysis of systems that can be described by these equations. You will learn that modeling systems containing two independent ideal energy storage devices always leads to a second-order linear differential equation. The material contained in this chapter, along with that contained in Chapters 3 and 4, is the foundation of any branch of engineering.

Chapter 7 is the second chapter in the book that relates theory to practical engineering. This time a real-world problem is selected that combines fluid, mechanical, and electrical systems modeling and analysis. You will see how everything you learned in the previous chapters is applied to solving complex engineering problems.

Chapter 8 is intended to lead you into the world of more complex engineering systems with the confidence that what you learned in the previous chapters is all that is required to understand such systems. You will learn that no matter how complex an engineering system is, it can be broken down into combinations of first- and second-order linear differential equations.

Appendix A provides an opportunity for you to review engineering calculus, for those who may need it. It covers the high points of both differential and integral calculus. You need to be familiar with at least the amount of calculus presented in this appendix to understand the rest of the material in this book.

Appendix B contains the physics behind the engineering building blocks covered in Chapter 2. At first I was going to put this information in Chapter 2, but it made for a very long chapter and I didn't want readers to get bogged down in all the math at the very beginning of the book.

The information in this appendix is very powerful and very useful, however, and I strongly recommend that you peruse and understand it.

I hope you enjoy the book and that it helps you achieve your goals in engineering.

Chapter

2 Basic Building Blocks for Modeling Engineering Systems

Objectives

At the completion of this chapter, you will be able to:

■ Define the basic concepts of voltage, current, work, power, and energy as related to electrical systems.

■ Identify the fundamental electrical circuit elements and write their describing equations.

■ Recognize two important tools used in formulating math models: impedance and operational block diagrams.

■ Define the basic concepts of motion and force as they relate to mechanical components.

■ Identify the fundamental mechanical components and write their describing equations.

■ Define the basic concepts of pressure and mass/volume rate of flow in fluid systems.

■ Identify the fundamental fluid components and write their describing equations.

■ Define the basic concepts of temperature and heat flow in thermal systems.

■ Identify the fundamental thermal elements and write their describing equations.

■ Recognize the analogies that can be drawn between the fundamental elements of all four types of systems: electrical, mechanical, fluid, and thermal.

2.1 Introduction

We'll now discuss how to build mathematical models of engineering systems. All of the material you will encounter in this chapter should be somewhat familiar, since it is covered in most first courses in physics. However, the subject matter is presented differently. If you don't understand something, I strongly suggest that you get a first-year college physics text and refer to it as required as you read though this chapter.

Throughout this book I deal primarily with *linear* equations because they are far easier to work with than nonlinear equations, and because they give you the quickest insight into the physical behavior of engineering systems. A linear equation is an equation in which a change in the input or independent variable results in a *proportional* change in the output or dependent variable. Take, for example, the equation for the volume of a cylinder V given by

$$V = \pi r^2 h \qquad (2.1)$$

where r = radius of the cylinder

   h = height of the cylinder.

This equation is linear with respect to h, but it is nonlinear with respect to r. You can see this by holding one variable constant while varying the other. For a cylinder where r is fixed, doubling h from h_0 to $2h_0$ doubles the volume of the cylinder, a proportional increase. Therefore the equation is linear with respect to h. For a cylinder where h is fixed, doubling the radius from r_0 to $2r_0$, does not result in a proportional increase in volume—it quadruples the volume. Therefore, the equation is nonlinear with respect to r.

Quite often you will want to linearize an equation so you can study the behavior of an engineering system model about a certain set of values for the independent variables. This set of values is frequently called the operating or steady-state point of the system. A very powerful equation called Taylor's Theorem allows any function of any number of independent variables to be expanded about an operating point. To fully understand and appreciate the power of this equation you need to understand the concept of partial derivatives. (See the review of these topics in Appendix A if you need a quick brush-up.)

Throughout the book, we'll go from the simplest elements to more and more complex systems. In this chapter, you will learn how to break down components found in electrical, mechanical, fluid, and thermal systems into rudimentary elements that can be described by simple differential and integral calculus. You will discover that there are only three fundamental elements in electrical, mechanical, and fluid systems, and only two in thermal systems. You will also discover that, from a mathematical point-of-view, all of the fundamental elements in each of these diverse fields look and behave exactly alike. You will learn that design and analysis problems in one field of engineering can be easily converted to another field, where other tools for obtaining solutions might be available.

In the interests of space and practicality, I've left out some of the physics in this chapter and placed it in Appendix B, *The Physics of Work, Power, and Energy in Engineering Systems*. In order to make sure that you understand the physics behind all of the elements described here, it would be a good idea to read through this appendix.

In this chapter, I'll begin a convention that I'll use throughout the book. Equations that I feel are important enough to commit to memory will be boxed in. You might wonder why you should commit *anything* to memory, when you could just look it up in a reference book when you need it. I strongly believe that a certain core set of equations should be memorized if you're going to be successful at engineering. These equations could be compared to a minimal vocabulary for someone who is learning to read. If we didn't remember the meanings of all the words we use in English, how would we read or speak? I suppose we could sit with a dictionary and look each word up, but that would be rather painful! If you understand how these equations are derived and how they relate to each other, your engineering practice will be smoother and more successful.

2.2 Electrical Elements

I'll begin this introduction with electrical systems, for the following reasons. First, every engineer and advanced technician, no matter what their field of specialty, should know everything contained in this section about electrical engineering, just on general principles! Second, many modeling and analysis concepts and techniques are easier to explain and easier to grasp using electrical elements. Third, numerous tools and methods for analyzing electrical circuits have been developed over the years and all of these can be applied to modeling and analyzing mechanical, fluid, and thermal circuits as well.

Unfortunately, I've discovered that many civil, mechanical, and other nonelectrical engineers and technicians shy away from electrical circuits, circuit modeling, circuit analysis, and other such nasties. *Please* don't do this. If you want to be a really good engineer or advanced technician, you should know all of the electrical engineering fundamentals discussed in this section.

Concepts of Voltage and Current

The concepts of *voltage* and *current* are used in electrical and electronics engineering to describe the behavior of engineering systems that use electricity. Electrical power supplies, generators, motors, transformers, and computers are examples of such systems.

The term *voltage* is used to describe the work that must be performed to move a unit of electrical *charge* (an electron is a minute electrical charge) from one point to another. Consequently, voltage is a relative term and is often referred to as the *potential difference* between two points. The units for voltage are *volts*, and the units for electrical charge are *coulombs*. One coulomb of electrical charge is equal, but opposite in sign, to 6.22×10^{18} electrons.

Figure 2.1 is a *definition sketch* which shows a symbolic, or circuit, diagram of an electrical element. A voltage V_1 is at one end and a voltage V_2 at the other. If V_2 is not equal to V_1, then electrical charge flows from one side of the element to the other. This flow of electrical charge per unit of time is called *current* and is given the symbol i. Current is measured in units called *amperes* or *amps*.

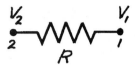

Figure 2.1. Definition sketch of electrical element.

Figure 2.2 shows the same electrical element given in Figure 2.1 with different voltage values at each end. This figure indicates that current has a direction and a magnitude. If the voltage at point 2 is greater than that at point 1, then by convention we say a current flows from point 2 to point 1. Conversely, the current flows from point 1 to point 2 if the voltage at point 1 is greater than it is at point 2. If the voltages at the two points are the same, no current flows.

Since voltage is a relative term, showing a voltage V_2 at point 2 and a voltage V_1 at point 1,

$V_2 > V_1$

$V_2 < V_1$

$V_2 = V_1$

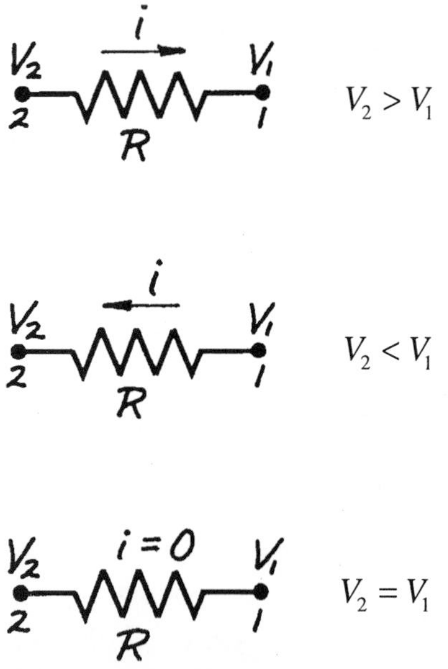

Figure 2.2. Definition sketch of electrical element with different voltage/current values.

as was done in Figures 2.1 and 2.2, implies that there is some reference point associated with these voltages. That point is generally called *ground* or *earth*. Ground has no potential; that is, there is no place surrounding ground where work must be done on an electrical charge to move it from that point to ground. The voltage at point 2 can be referenced to the voltage at point 1. We will use the symbol V_{21}, meaning $V_{21} = V_2 - V_1$, to denote this reference.

Concepts of Work, Power, and Energy in Electrical Elements

Since voltage is defined as the work that must be done to move a unit of electrical charge from one point to another, we can write voltage between two points as

$$V_{21} = \frac{dW_{21}}{dq} \qquad (2.2)$$

where dW_{21} is the work that must be done to move the electrical charge dq from point 1 to point 2. A unit of measure for work is the *joule*. Equation (2.2) then defines volts as joules per coulomb. That is, one volt is equal to one joule of work per one coulomb of charge. One joule of work is equivalent to one watt-sec or 0.737 ft-lbs.

Current was defined above as the flow of electrical charge per unit of time. This can be written in the form of a derivative. That is,

$$i = \frac{dq}{dt} \qquad (2.3)$$

The units which apply to this equation are coulombs per second, called amperes.

Power is defined as the rate at which work is performed. This too can be written in the form of a derivative. That is,

$$P = \frac{dW}{dt} \qquad (2.4)$$

The product of voltage differential across an electrical element and the current flowing through the element is equal to power. This is a very important concept. You can see this by multiplying equation (2.2) and (2.3). That is,

$$V_{21} \times i = \frac{dW_{21}}{dq} \times \frac{dq}{dt} = \frac{dW_{21}}{dt} = P \qquad (2.5)$$

The units for power are

$$P = V_{21}i = 1 \text{ volt} \times 1 \text{ ampere}$$

or

$$P = V_{21}i = 1\frac{\text{joule}}{\text{coulomb}} \times 1\frac{\text{coulomb}}{\text{second}}$$

$$= \frac{\text{joules}}{\text{second}} = 1 \text{ watt}$$

Since work is a transitory form of energy, we can rewrite equation (2.4) as

$$\frac{dE}{dt} = P \qquad (2.6)$$

Equation (2.6) can be integrated to obtain the energy stored in, or dissipated by, an electrical element over a time interval from $t = t_a$ to $t = t_b$. That is,

$$dE = Pdt$$

$$E = \int Pdt$$

$$E_b - E_a = \int_{t_a}^{t_b} Pdt = \int_{t_a}^{t_b}(V_{21}i)dt \qquad (2.7)$$

The Resistor

The most common of all electrical elements is the resistor. It is intentionally or unintentionally present in every real electrical system. Figure 2.3 shows a symbolic (circuit) diagram and a graphical representation of this element. Also shown are the fundamental describing equations for an *ideal resistor*. An idealization of the resistor is given by

$$V_{21} = Ri \qquad (2.8)$$

Ri is a linear function in which the voltage is proportional to the current and R is the constant of proportionality. You will also often see the equation for a resistor written as

$$i = \frac{V_{21}}{R} \qquad (2.9)$$

Equations (2.8) and (2.9) are often called *Ohm's Law*. The value of R is usually given in *ohms*, which is really volts per amp.

The energy delivered to a resistor in the interval from $t = t_a$ to $t = t_b$ is given by equation (2.7). We can substitute equation (2.9) into (2.7) and eliminate the current as follows:

$$E_b - E_a = \int_{t_a}^{t_b}(V_{21}i)dt \qquad (2.7)$$
repeated

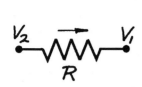

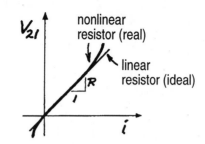

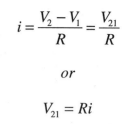

(a). Symbolic or circuit diagram.	(b). Graphical representation.	(c). Describing equations for ideal damper.

Figure 2.3. The resistor.

Substitute V_{21} / R for i and obtain

$$E_b - E_a = \int_{t_a}^{t_b} V_{21}\left(\frac{V_{21}}{R}\right) dt \qquad (2.10)$$

Since R is a constant, it can be pulled outside of the integral sign, giving

$$E_b - E_a = \frac{1}{R}\int_{t_a}^{t_b} V_{21}^2 dt \qquad (2.11)$$

We could have substituted equation (2.8) into equation (2.7) and eliminated V_{21}. If we take this route then we get

$$E_b - E_a = \int_{t_a}^{t_b} (Ri) i\, dt \qquad (2.12)$$

Pulling the constant R outside the integral sign gives

$$E_b - E_a = R\int_{t_a}^{t_b} i^2 dt \qquad (2.13)$$

One very important thing that equations (2.11) and (2.13) reveal is that a resistor dissipates power. Regardless of the direction of the current or the sign of the voltage, both are squared in the equation. Therefore, energy can't

be retrieved from a resistor. This element only dissipates energy.

By comparing equation (2.6) with (2.11) and (2.13) you can see that the power dissipated by a resistor at any instant in time is

$$P = \frac{V_{21}^2}{R} = i^2 R \qquad (2.14)$$

In real electrical circuits, the energy dissipated by a resistor is converted into heat. Unless this heat is removed the resistor could burn out.

The Capacitor

Another fundamental electrical element is the *capacitor*. Like the resistor, it is intentionally or unintentionally present in every real electrical system. Figure 2.4 shows a symbolic (circuit) diagram and a graphical representation of this element. Also shown are the fundamental describing equations for an *ideal capacitor*.

A capacitor is constructed of two pieces of conducting material separated by another material that allows an electrostatic field to be established without allowing charge to flow between

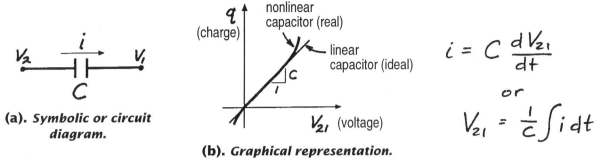

Figure 2.4. The capacitor.

(a). *Symbolic or circuit diagram.*

(b). *Graphical representation.*

(c). *Describing equations for ideal capacitor.*

the two pieces of conducting material. A capacitor stores electrical energy in this electrostatic field. In an ideal capacitor, all of the energy stored in the device can be retrieved and used.

An idealization of the capacitor is given by the linear relationship

$$q = CV_{21} \qquad (2.15)$$

You can see from equation (2.15) that units of capacitance are coulombs per volt. One coulomb per volt is called a *farad*. A farad is a very large number, so most capacitors you will run into will have values given in microfarads (μf) which is one-millionth of a farad.

Differentiating both sides of equation (2.15) with respect to t gives

$$\frac{dq}{dt} = C\frac{dV_{21}}{dt} \qquad (2.16)$$

Combining (2.16) and (2.3) gives the following equation for the current-voltage relationship of an ideal capacitor

$$i = C\frac{dV_{21}}{dt} \qquad (2.17)$$

The energy delivered to a capacitor in the time interval from $t = t_a$ to $t = t_b$ is from equation (2.7) given by

$$E_b - E_a = \int_{t_a}^{t_b} V_{21} i \, dt \qquad (2.18)$$

substituting CdV_{21} for $i \, dt$ from equation (2.17) gives

$$E_b - E_a = C\int_{V_a}^{V_b} V_{21} dV_{21} \qquad (2.19)$$

Note that the limits of integration have been changed. V_b is the voltage across the capacitor at time $t = t_b$ and V_a is the voltage at time $t = t_a$. The equation can be integrated to give

$$E_b - E_a = C\left[\frac{V_{21}^2}{2}\right]_{V_a}^{V_b} = C\frac{V_b^2}{2} - C\frac{V_a^2}{2} \qquad (2.20)$$

The quantity $C(V_a^2/2)$ is the energy that was initially stored in the capacitor at $t = t_a$ and $C(V_b^2/2)$ represents the energy stored at time $t = t_b$. During the time interval $t_b - t_a$, the energy $(E_b - E_a)$ was added to the capacitor. The energy storage feature of a capacitor is extremely important. Numerous electrical circuits, including memory chips in computers make use of the energy storage capability of a capacitor.

The Inductor

The third and last electrical element we will discuss is the inductor. Like the resistor and capacitor, it is intentionally or unintentionally present in every real electrical system. Figure 2.5 shows a symbolic (circuit) diagram and a graphical representation of this element. Also shown are the fundamental describing equations for an *ideal inductor*. It is essentially the opposite of a capacitor. The symbol for an inductor looks like a wire coil because, in its simplest form, that's all it is. Any wire or conductor carrying a current is surrounded by a magnetic field. This magnetic field can be concentrated by winding the wire into a tight coil about a tube. The strength of the magnetic field due to one turn φ can be increased by increasing the number of turns N, so the strength of the total magnetic field is given by $N\varphi$. The magnetic field strength can be increased even further by wrapping coils of the wire around an iron core. However, this generally makes the inductor nonlinear and creates hysteresis, as shown in the graph of Figure 2.5.

The strength of a magnetic field around and the voltage across an inductor are related by the following equation, often referred to as *Faraday's law of induction*:

$$V_{21} = \frac{d(N\varphi)}{dt} \qquad (2.21)$$

If no magnetic materials are present or used in the core of the inductor, then the strength of the magnetic field is linearly dependent on the current. That is,

$$\frac{d(N\varphi)}{dt} = L\frac{di}{dt} \qquad (2.22)$$

Combining equations (2.21) and (2.22) gives

$$V_{21} = L\frac{di}{dt} \qquad (2.23)$$

Note that the units of an inductor as determined from equation (2.23) are volts times seconds per ampere. One volt-second per ampere is called a *henry* (H). Like the farad, a henry is a large unit. Consequently, most real-life induc-

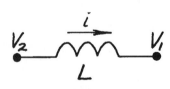

(a). Symbolic or circuit diagram.

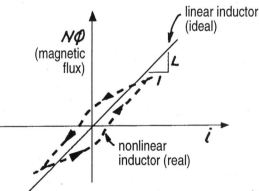

linear inductor (ideal)

$N\varphi$
(magnetic flux)

L

nonlinear inductor (real)

i

$i = \frac{1}{L}\int V_{21}\,dt$

or

$V_{21} = L\frac{di}{dt}$

(b). Graphical representation.

(c). Describing equations for ideal inductor.

Figure 2.5. The inductor.

tors you will run into will have values in the millihenry (mH) or microhenry (µH) range. A millihenry is one-thousandth of a henry and a microhenry is one-millionth of a henry.

The energy delivered to an inductor in the time interval $t = t_a$ to $t = t_b$ is

$$E_b - E_a = \int_{t_a}^{t_b} V_{21} i dt \qquad \text{(2.7)}$$
repeated

Substituting equation (2.23) for V_{21} gives

$$E_b - E_a = \int_{t_a}^{t_b} \left(L \frac{di}{dt} \right) i dt = L \int_{i_a}^{i_b} i di$$

Carrying out the integration gives

$$E_b - E_a = L \left[\frac{i^2}{2} \right]_{i_a}^{i_b} = L \frac{i_b^2}{2} - L \frac{i_a^2}{2} \quad \text{(2.24)}$$

The quantity $L(i_a^2 / 2)$ is the energy that was initially stored in the inductor (magnetic field) at $t = t_a$ and $L(i_b^2 / 2)$ is the energy stored at $t = t_b$. During the time interval $t_b - t_a$, the energy $(E_b - E_b)$ was added to the magnetic field of the inductor. The energy storage feature of an inductor is extremely important. Numerous electrical circuits including electromagnets, radios, and switching power supplies make use of it.

What to Commit to Memory

Figure 2.6 shows the three fundamental electrical elements and one describing equation for each. You should commit these equations to memory. Also commit to memory the fact that power is the product of voltage and current. If

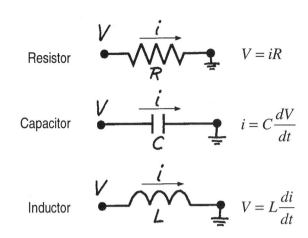

Figure 2.6. The three fundamental electrical elements.

you remember all this, you will be able at any time to derive all of the energy equations discussed here and used later on in the book.

Impedance and Operational Block Diagrams

Now that you know the equations that describe the relations between voltage and current for ideal resistors, capacitors, and inductors, I want to introduce some very important tools that will help you remember these equations and that will make it easier to derive mathematical models of systems containing the basic electrical components. You will discover in later chapters that these tools are applicable not only to electrical systems, but also to mechanical, fluid, and thermal systems.

Tool #1 – Impedance

The concept of impedance is based on observing that a variable which flows through an element is impeded by the element. For example, in electrical elements a current flows *through* a resistor due to a voltage difference *across* the resistor. The current is called the *through* variable and voltage is called the *across* variable. These terms apply to inductors and capacitors also.

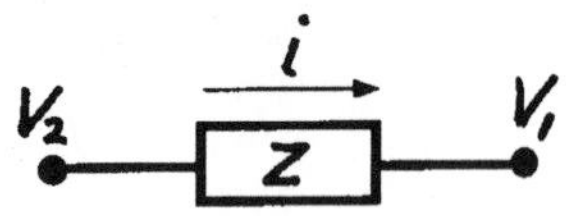

Figure 2.7. Representation of the imped-ance Z of any electrical element.

The impedance of any electrical component, shown in Figure 2.7, is given by

$$Z = \frac{V_{21}}{i} \qquad (2.25)$$

For a resistor, the impedance Z_R is given by

$$Z_R = \frac{V_{21}}{i} = R \qquad (2.26)$$

That is, the impedance of a resistor is simply its resistance.

The relationship between voltage and current for a capacitor is given by

$$V_{21} = \frac{1}{C}\int i\,dt \qquad (2.27)$$

I will often use the operator $D = d/dt$ to express derivatives (and $1/D$ to express integrals). This notation helps to reduce the derivation of math models to mere multiplication. (It's discussed in more detail in Appendix A.) Using the operator notation, we can write this equation as

$$V_{21} = \frac{1}{CD}i \qquad (2.28)$$

The impedance is then given by

$$Z_C = \frac{V_{21}}{i} = \frac{1}{CD} \qquad (2.29)$$

The relationship between voltage and current for an inductor is given by

$$V_{21} = L\frac{di}{dt} \qquad (2.30)$$

Again using the operator notation, we can write this equation as

$$V_{21} = LDi \qquad (2.31)$$

The impedance is then given by

$$Z_L = \frac{V_{21}}{i} = LD \qquad (2.32)$$

Figure 2.8 provides a summary of the imped-ance of electrical circuit elements.

Resistor Impedance

Capacitor Impedance

Inductor Impedance

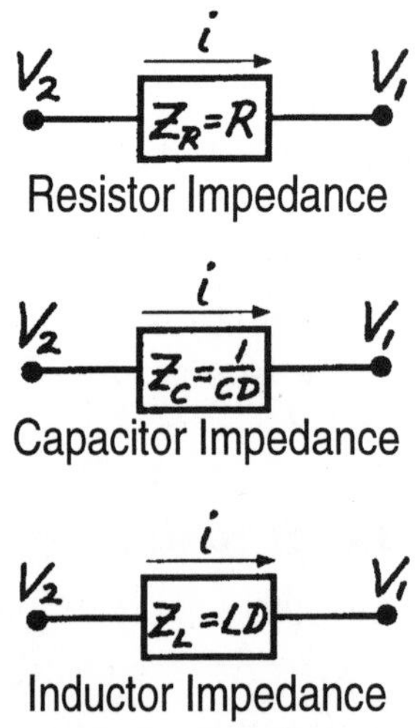

Figure 2.8. Impedance of electrical circuit elements.

Tool #2 – Operational Block Diagrams

Operational block diagrams are extremely helpful in visualizing an engineering system and in communicating your design ideas to others. The completed diagram can also assist with solving the equations using analog and digital computers.

An operational block represents a mathematical operation. It operates on the input or forcing variable and transforms it into the output or response variable. An operational block is often called a transfer function. That is it transfers (transforms) the input variable into the output variable and clearly establishes cause and effect in a system.

There are four basic operational blocks:
(1) the summer
(2) the constant multiplier
(3) the integrator
(4) the differentiator.

An example of a summer is shown in Figure 2.9. There are two input signals in this example, V_1 and V_2. Input signals are designated as such by showing an arrow pointing into the summation box along with a plus or minus sign. There can only be one output from a summer. In this case it is V_3.

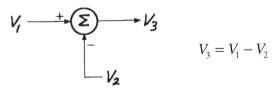

$$V_3 = V_1 - V_2$$

(a). *Block diagram.* **(b).** *Equation.*

Figure 2.9. The operational block diagram of a summer.

An example of a constant multiplier block is shown in Figure 2.10. I use the notation that the variable on the right side of an equal sign in an equation is the input or forcing variable, and the variable on the left side is the output or response variable. The block diagram makes this very clear.

$$i \longrightarrow \boxed{R} \longrightarrow V_{21} \qquad V_{21} = Ri$$

(a). *Block diagram.* **(b).** *Equation.*

Figure 2.10. The operational block diagram of a multiplier.

Even though we know we can solve for i in terms of V_{21}, the block diagram does not permit this. You must rewrite the equation and draw a new block diagram as shown in Figure 2.11.

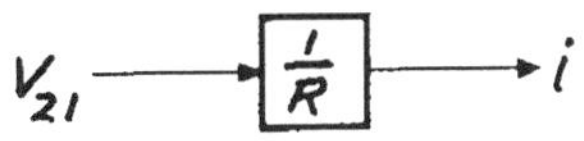

$$V_{21} \longrightarrow \boxed{\frac{1}{R}} \longrightarrow i \qquad i = \frac{1}{R}V_{21}$$

(a). *Block diagram.* **(b).** *Equation.*

Figure 2.11. The operational block diagram of a multiplier showing importance of distinguishing between input and output.

The next block diagram we will discuss is the integrator. As you know, integration must account for the *constant of integration*. To handle this in a block diagram, a summer is added *after* the integration block, as shown in Figure 2.12. Be careful here. The summer and the integration

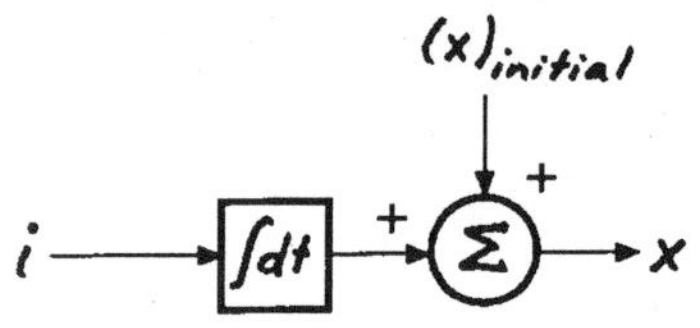

(a). Block diagram.   **(b). Equation.**

$$x = \int_o^t i\,dt + (x)_{init}$$

Figure 2.12. The operational block diagram of an integrator.

(a). Block diagrams.   **(b). Equations.**

$$x_1 = \frac{dx_2}{dt}$$

$$x_1 = Dx_2$$

Figure 2.14. Operational block diagrams of a differentiator.

block go together, and the summer must be *after* the integrator. If it were in front and had a value, it would produce an incorrect answer.

I used the integral sign in Figure 2.12 to remind you that it is equivalent to the 1/D operator. An equivalent integrator block diagram is shown in Figure 2.13.

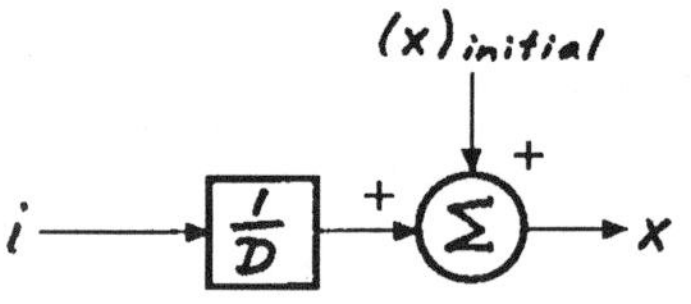

(a). Block diagram.   **(b). Equation.**

$$x = \frac{1}{D}i + (x)_{initial}$$

Figure 2.13. Alternative operational block diagram for an integrator.

The final block diagram we'll discuss is the differentiator. An example is shown in Figure 2.14. A differentiator can sometimes be useful, but great care must be taken in their use. For example, I've designed a number of instrumentation systems that measure ship motions and accelerations. It is possible to directly measure only displacement and then differentiate to get velocity and acceleration. However, since differentiators tend to magnify noise and are inaccurate on a digital computer compared to

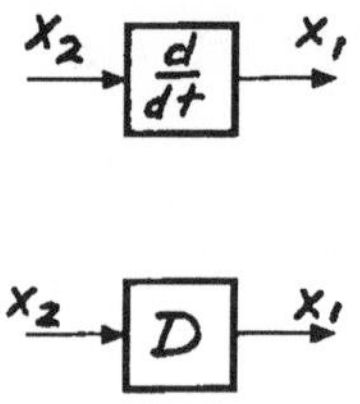

integration, I usually use accelerometers to measure acceleration directly. Integration, on the other hand, is a smoothing operation and can be handled accurately on a digital computer. You should typically avoid using differentiators in your block diagrams for the three basic electrical components.

We will use these tools extensively throughout this book. Figure 2.15 provides a summary of the three electrical circuit elements. It shows the symbolic circuit diagram, the operational block diagrams, and the impedance circuit.

2.3 Mechanical Components (Translational)

Concepts of Mechanical Motions and Forces

Motion and *force* are concepts used to describe the behavior of engineering systems that employ mechanical components. Heads of disk drives, armatures of motors, gears, ships, and automobiles are just a few examples of engineering systems that employ mechanical components. Indeed, it is hard to imagine any engineering system that does not have mechanical components.

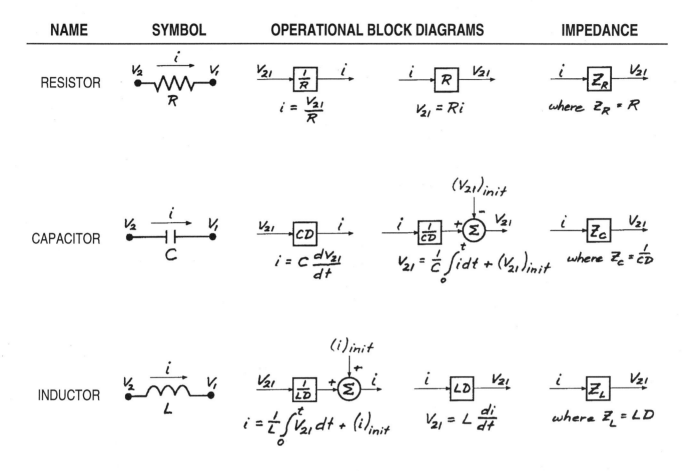

NAME	SYMBOL	OPERATIONAL BLOCK DIAGRAMS		IMPEDANCE

Figure 2.15. Summary of fundamental electrical elements.

Motion is a term used to describe the movement of a point relative to another, and it is described using the terms *distance*, *velocity*, and *acceleration* (see Appendix A if you need a brush-up). The three are related by differentiation or integration. If you know one, you can obtain the other two.

Figure 2.16 is a symbolic or "circuit" diagram of a mechanical component whose ends are undergoing translational movement. One end of this component is moving in a straight line at a

velocity v_2 and the other is moving in the same straight line at a velocity v_1. Since velocity is a relative term, this figure implies the existence of a reference that is fixed.

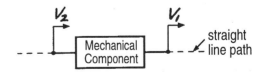

Figure 2.16. Symbolic diagram of a translational mechanical component.

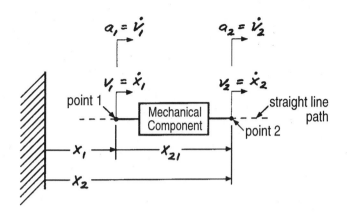

Figure 2.17. Symbolic diagram of a translational mechanical component showing reference point.

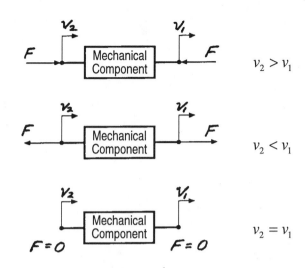

Figure 2.18. Symbolic diagram of a translational mechanical component showing direction and magnitude of force.

Figure 2.17 shows the same mechanical component as in Figure 2.16 but with the fixed reference shown. The component is moving in a straight line and I've shown the distance, velocity, and acceleration of points 1 and 2.

From this figure we can write the following equations describing the motion of point 2 relative to point 1:

$$x_{21} = x_2 - x_1 \qquad (2.33)$$

$$v_{21} = v_2 - v_1 = \dot{x}_2 - \dot{x}_1 = \dot{x}_{21} \qquad (2.34)$$

$$a_{21} = a_2 - a_1 = \dot{v}_2 - \dot{v}_1$$
$$= \dot{v}_{21} = \ddot{x}_2 - \ddot{x}_1 = \ddot{x}_{21} \qquad (2.35)$$

Relative motion between the ends of a mechanical component can't exist without a force being present. When relative motion exists, the mechanical component is placed in a state of tension or compression. The associated force has both a magnitude and a sign. The conventions used are shown in Figure 2.18.

Note that power is the product of the force and velocity:

$$Power = Fv$$

Refer to Appendix B.2 for more details on work, energy, and power.

The Damper or Dashpot

A damper (sometimes called a dashpot) is a mechanical component often found in engineering systems. The shock absorbers in your car are an example of a mechanical damper that is intentionally designed into every car. A typical shock absorber is shown in Figure 2.19. The device consists of a piston that moves inside a cylinder filled with hydraulic fluid. Small holes are drilled through the piston so fluid can move from one chamber to the other. As the piston moves relative to the cylinder, the fluid is forced through these small openings, creating resisting fluid shearing forces. If the mass and springiness

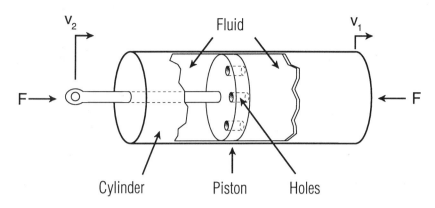

Figure 2.19. A typical mechanical damper.

An ideal damper is a linear component described by

$$F = b(v_2 - v_1) = bv_{21}$$

(2.36)

or

$$v_{21} = \frac{1}{b}F$$

(2.37)

of the piston and cylinder are small, then the force will be a function of the relative velocity between the piston and the cylinder.

Figure 2.20 shows a symbolic (circuit) diagram of this component along with a graphical representation and the fundamental describing equations for an ideal damper. The symbol implies that this component has no mass and the connecting rods have no springiness. Furthermore, because this is a translational component, the forces act along the same single straight line that characterizes the motion of the two ends.

Knowing the units of F and v, you can see that the units of the damping force constant b are lb-sec/in.

Like the electrical resistor, the energy delivered to a mechanical damper cannot be retrieved. This energy is dissipated to the surroundings in the form of heat. (Refer to Appendix B for the derivation equations.)

The power dissipated by the damper is

$$P = F \times v_{21} = (bv_{21}) \times v_{21}$$
$$= bv_{21}^2$$

(2.38)

or

$$P = F \times v_{21} = F \times \left(\frac{1}{b}F\right)$$
$$= \frac{1}{b}F^2$$

(2.39)

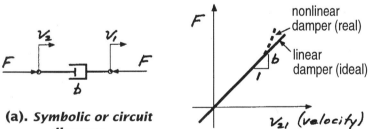

(a). Symbolic or circuit diagram.

(b). Graphical representation.

$$F = b(v_2 - v_1) = bv_{21} \quad \text{OR} \quad v_{21} = \frac{1}{b}F$$

(c). Describing equations for ideal damper.

Figure 2.20. The ideal damper.

The Translational Mass

All real mechanical components used in engineering systems have mass. In general, the mass is distributed in three-dimensional space. However, it is frequently possible (and very convenient) to treat a component as if all of its mass were concentrated at a single point called the center of gravity, or c.g. That is, we lump the mass together into one point and end up with an ideal, *lumped parameter* component. If we further consider only a lumped parameter mass moving in a straight line, then we have a *translational mass*.

The symbolic diagram that will be used for a translational mass component is a little peculiar. So, let's build up to it slowly. Figure 2.21 shows a lumped mass moving in a straight line. According to Newton, if the mass m is not changing then

$$F = ma_2 = m\frac{dv_2}{dt} = m\frac{d^2x_2}{dt^2} \qquad (2.40)$$

Now let's write equation (2.40) in terms of relative velocity between points 2 and 1

$$F = m\frac{dv_{21}}{dt} \qquad (2.41)$$

and then determine what conditions we have to place on point 1 for equation (2.41) to apply. If we differentiate equation (2.34) we have

$$\frac{dv_{21}}{dt} = \frac{dv_2}{dt} - \frac{dv_1}{dt} \qquad (2.42)$$

Now if we combine equations (2.41) and (2.42) we get

$$F = m\frac{dv_2}{dt} - m\frac{dv_1}{dt} \qquad (2.43)$$

Finally, compare equation (2.43) with equation (2.40). The two equations are equal only when v_1 is constant. In that case dv_1/dt will equal zero. This means that point 1, the reference, must be stationary or moving at a constant velocity in order for equation (2.41) to be true.

The symbolic diagram we will use to represent the translational mass component is given in Figure 2.22. The dotted line at one end of the element will serve to remind us that

(a) there is no physical connection between point 2 (the mass) and point 1

(b) point 1 must be stationary or moving at a constant velocity.

(c) no force is transmitted to point 1.

Figure 2.23 shows the symbolic (circuit) diagram of the translational mass along with a graphical representation and the fundamental describing equations. Note that the graph shows a plot of *momentum* versus

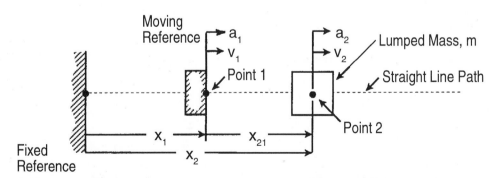

Figure 2.21. Lumped mass moving in a straight line.

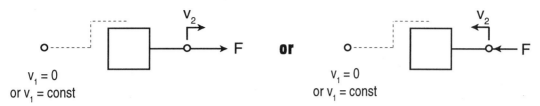

Figure 2.22. Symbolic diagram for translational mass.

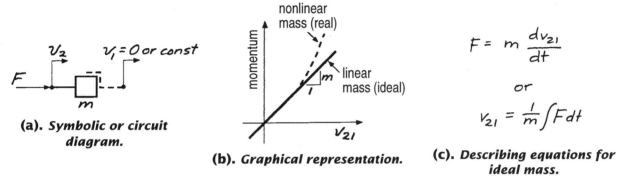

(a). Symbolic or circuit diagram.

(b). Graphical representation.

(c). Describing equations for ideal mass.

Figure 2.23. The ideal translational mass.

velocity. Momentum is the product of the mass of a moving object and its velocity. Force is equal to the rate of change of momentum. That is

$$F = \frac{d(mv)}{dt} \qquad (2.44)$$

Only when m is constant can we pull the m outside of the differential. If m is constant then the mass is called an ideal (or Newtonian) mass. That is,

$$F = m\frac{dv_{21}}{dt} \qquad (2.45)$$

or

$$v_{21} = \frac{1}{m}\int F\,dt \qquad (2.46)$$

Knowing the units of F and v, you can always determine the units of mass as $F/(dv/dt) =$ lb-sec²/in.

The energy delivered to the mass in the time interval $t = t_a$ to $t = t_b$ by a force $F(t)$ is

$$E_b - E_a = \int_{t_a}^{t_b} F v_{21}\,dt \qquad (2.47)$$

Substituting (2.45) into (2.47) to eliminate F gives

$$E_b - E_a = \int_{t_a}^{t_b}\left(m\frac{dv_{21}}{dt}\right)v_{21}\,dt$$

$$= m\int_{v_{21a}}^{v_{21b}} v_{21}\,dv_{21} \qquad (2.48)$$

Carrying out the integration gives

$$E_b - E_a = m\left[\frac{v_{21}^2}{2}\right]_{v_{21a}}^{v_{21b}} = \frac{m}{2}\left(v_{21b}^2 - v_{21a}^2\right)$$

$$= \frac{m}{2}v_{21b}^2 - \frac{m}{2}v_{21a}^2 \qquad (2.49)$$

Let's examine equation (2.49). The quantity $(m/2)v_{21a}^2$ represents the energy of the mass at time $t = t_a$ and the quantity $(m/2)v_{21b}^2$ represents the energy at time $t = t_b$. These quantities are identical to those of the capacitor. Like the capacitor, the mass is an energy storage device. The energy stored can be completely retrieved.

The Translational Spring

Nearly all materials used in mechanical systems exhibit an elastic effect that we call a spring. When a force is applied to these materials they deform. When the force is removed, they return to their original shape. A spring is a fundamental mechanical component found intentionally or unintentionally in almost every mechanical engineering system. The springs in your car are a good example. Steel is shaped into a coil so the mass of the spring is minimized and its elasticity maximized.

Figure 2.24 shows a symbolic (circuit) diagram of a spring along with a graphical representation and the fundamental describing equations. The symbol implies that this component has no mass or damping. Furthermore, because this is a translational component, the forces act along the same single straight line that characterizes the motion of the two ends.

An ideal spring is described by

$$F = kx_{21} \qquad (2.50)$$

where k is a constant of proportionality relating the force F to the deformation x_{21} of the spring. Knowing the units of F and x, you can easily determine the units of k as $F/x = $ lb/in.

Since we want to describe all of the components in terms of velocity and force, we'll differentiate equation (2.50) as follows

$$\frac{dF}{dt} = k\frac{dx_{21}}{dt} = kv_{21} \qquad (2.51)$$

Rearranging gives

$$\boxed{v_{21} = \frac{1}{k}\frac{dF}{dt}} \qquad (2.52)$$

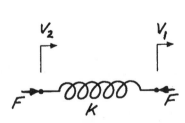

(a). **Symbolic or circuit diagram.**

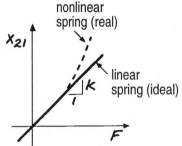

(b). **Graphical representation.**

$$F = k\int v_{21}\,dt$$

or

$$v_{21} = \frac{1}{k}\frac{dF}{dt}$$

(c). **Describing equations for ideal spring.**

Figure 2.24. The ideal spring.

Also we can integrate (2.51) as follows

$$F = \int dF = \int k v_{21} dt$$

or

$$\boxed{F = k \int v_{21} dt} \qquad (2.53)$$

The energy delivered to the spring in the time interval $t = t_a$ to $t = t_b$ by a force $F(t)$ is

$$E_b - E_a = \int_{t_a}^{t_b} F v_{21} dt \qquad (2.54)$$

Substituting (2.52) into (2.54) to eliminate v gives

$$E_b - E_a = \int_{t_a}^{t_b} F\left(\frac{1}{k}\frac{dF}{dt}\right)dt$$

$$= \frac{1}{k}\int_{F_a}^{F_b} F\, dF \qquad (2.55)$$

Carrying out the integration gives

$$E_b - E_a = \frac{1}{k}\left[\frac{F^2}{2}\right]_{F_a}^{F_b}$$

$$= \frac{F_b^2}{2k} - \frac{F_a^2}{2k} \qquad (2.56)$$

Let's examine equation (2.56). The quantity $F_a^2/2k$ represents the energy of the spring at time $t = t_a$ and the quantity $F_b^2/2k$ represents the energy at time $t = t_b$. These equations are identical to those of the inductor. Like the inductor, the spring is an energy storage device. The energy stored can be completely retrieved.

What to Commit to Memory

You now have all the building blocks you need to mathematically model engineering systems comprised of translational mechanical components. In the next section we will look at rotational mechanical components, so we can model *any* mechanical system. However, before you move on, the following basic equations should be committed to memory:

DASHPOT

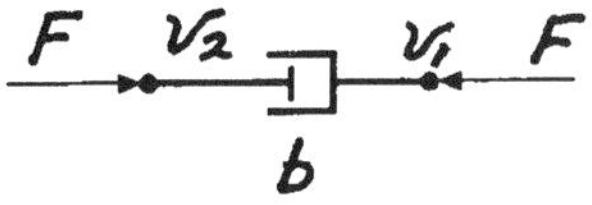

$$F = b v_{21}$$

MASS

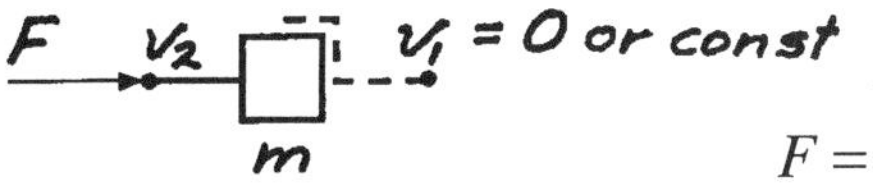

$$F = m\frac{dv_{21}}{dt}$$

SPRING

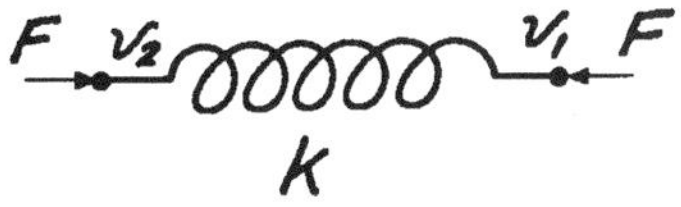

$$F = k x_{21} \quad \text{OR} \quad \frac{dF}{dt} = k v_{21}$$

Figure 2.25. Translational mechanical elements.

Just as I asked you to do with the basic electrical components, make these equations a part of your life and know them as well as you know your name!

The tools associated with operational block diagrams and impedance that you were introduced to when we investigated the basic electrical components also apply to the three basic translational mechanical components. You should now review **What to Commit to Memory** in the previous section on electrical elements. The block diagrams and impedances for the basic translational mechanical elements are given in Figure 2.26.

Analogies and Similarities

It has probably occurred to you by now that there are a lot of similarities between the equations describing mechanical and electrical components. Let's look into this in more detail. In Table 2.1 I have arranged the equations in such a way that it is obvious that they have the same form. If we were to agree that:

(1) voltage in an electrical system is analogous to velocity in a mechanical system, and

(2) current in an electrical system is analogous to force in a mechanical system,

then the electrical and mechanical components are *analogs* of one another. Resistors behave like dampers, capacitors like masses, and inductors like springs.

We could, however, just as easily arrange the equations as shown in Table 2.2. Now, if we were to agree that:

(1) voltage is analogous to force, and
(2) current is analogous to velocity

then we have again made electrical and mechanical components analogs of one another.

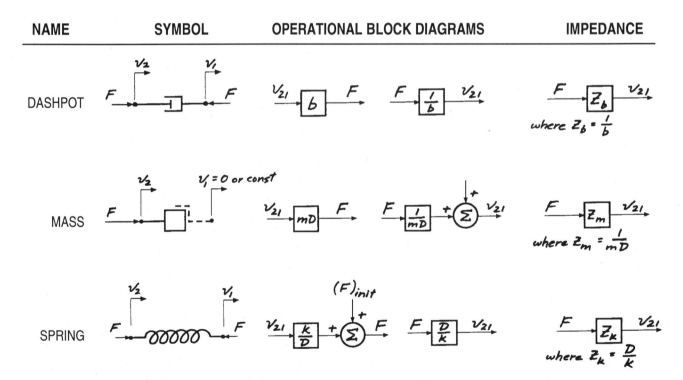

Figure 2.26. Summary of translational mechanical elements.

Table 2.1. Electrical–Mechanical Analogies – Version 1

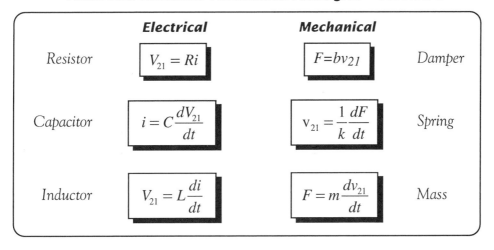

	Electrical	Mechanical	
Resistor	$V_{21} = Ri$	$v_{21} = \dfrac{1}{b}F$	Damper
Capacitor	$i = C\dfrac{dV_{21}}{dt}$	$F = m\dfrac{dv_{21}}{dt}$	Mass
Inductor	$V_{21} = L\dfrac{di}{dt}$	$v_{21} = \dfrac{1}{k}\dfrac{dF}{dt}$	Spring

Table 2.2. Electrical–Mechanical Analogies – Version 2

	Electrical	Mechanical	
Resistor	$V_{21} = Ri$	$F = bv_{21}$	Damper
Capacitor	$i = C\dfrac{dV_{21}}{dt}$	$v_{21} = \dfrac{1}{k}\dfrac{dF}{dt}$	Spring
Inductor	$V_{21} = L\dfrac{di}{dt}$	$F = m\dfrac{dv_{21}}{dt}$	Mass

I cannot overemphasize the importance of these mathematical analogies! In my opinion, they are one of the wonders of the world, and a terrific tool that can make your engineering design life easier. These analogs allow you to "see" and "feel" the behavior of all systems, regardless of your engineering background.

Which analogy is "correct"? Both! It really doesn't matter which one you choose, as long as you are consistent. I prefer the voltage–velocity and current–force analogy (Table 2.1), because it keeps the concepts of *across* and *through* variables the same in both electrical and mechanical engineering fields. That can be important when drawing schematic diagrams of systems and using the impedance tools introduced earlier. I'll be using the analogies that maintain the across and through variable relationships (i.e., those in Table 2.1) throughout this book.

2.4 Mechanical Components (Rotational)

Concepts of Angular Motions and Torques

Angular motion and *torque* are concepts used to describe the behavior of a certain class of mechanical components that undergo rotation. If an axis can be found about which all particles in a mechanical element rotate, then pure rotational motion exists. Motor armatures, ship propellers, and engine drive shafts are just a few examples of mechanical components in pure rotation.

Angular motion is a simple extension of linear motion and is easy to understand. Figure 2.27 shows a rotating shaft. The angular displacement θ, angular velocity ω, and angular acceleration α are all similar to linear displacement, velocity, and acceleration. However, instead of distance as a unit of measurement, we use an arc of a circle, called a radian or degree, to measure angular displacement. There are 2π radians and 360 degrees in a circle, so we can express angular velocity in radians per second or degrees per second. Similarly, we can express

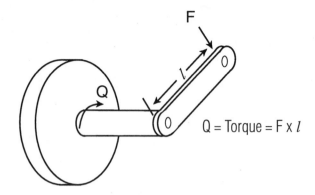

Figure 2.28. Rotating shaft with lever attached.

$$Q = \text{Torque} = F \times l$$

angular acceleration in radians per second per second or degrees per second per second.

Torque is to rotational mechanical components as force is to translational components. Torque is best viewed as a force acting on a lever arm. For example, Figure 2.28 shows a lever attached to the end of the shaft shown in Figure 2.27. A force F is applied a distance l away from the center of the shaft. A torque Q, equal to $F \times l$, is applied to the shaft. The units of torque are force times length or in-lbs.

As we did with translational mechanical elements, we can now visualize a generalized rotational mechanical element, as in Figure 2.29, and write the relative motion equations

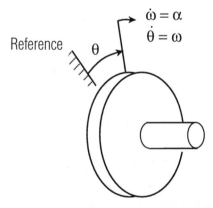

Figure 2.27. Rotating shaft.

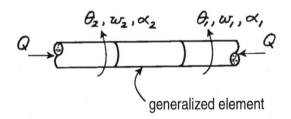

Figure 2.29. Definition sketch for generalized rotational mechanical element.

$$\theta_{21} = \theta_2 - \theta_1 \qquad (2.57)$$

$$\omega_{21} = \dot{\theta}_{21} = \dot{\theta}_2 - \dot{\theta}_1 = \omega_2 - \omega_1 \qquad (2.58)$$

$$\alpha_{21} = \dot{\omega}_{21} = \dot{\omega}_2 - \dot{\omega}_1$$
$$= \ddot{\theta}_2 - \ddot{\theta}_1 = \ddot{\theta}_{21} \qquad (2.59)$$

For rotational mechanical components, power is expressed as

$$Power = Q\omega_{21} \qquad (2.60)$$

Refer to Appendix B.3 for more details.

The Basic Rotational Mechanical Components

As you might expect, the same three fundamental mechanical elements, the damper, mass, and spring that we discussed in the translational mechanical section, also exist as rotational mechanical elements. We will therefore pass through these quickly, pointing out significant differences when appropriate.

The rotational mechanical damper symbolic diagram is shown in Figure 2.30. The torque is given by

$$Q = B(\omega_2 - \omega_1) = B\omega_{21} \qquad (2.61)$$

This equation is essentially identical to the one for the translational damper. The capital B is used to distinguish the rotational damper value symbol from that used with the translational damper. B has units of torque per unit of angular velocity, or in-lb-sec.

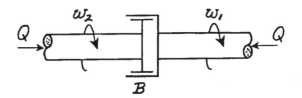

Figure 2.30. Rotational damper/dashpot.

The rotational mechanical mass symbolic diagram is shown in Figure 2.31. Newton's laws of motion applied to a pure rotational mass result in

$$Q = I\frac{d\omega_{21}}{dt} \qquad (2.62)$$

This equation is essentially identical to the one for the translational mass. The same restrictions apply to the reference angular velocity as were noted for the translational mass; that is, the reference point for angular velocity must be either stationary or moving at a constant angular velocity.

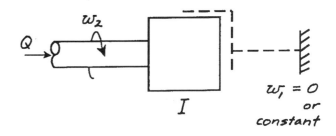

Figure 2.31. Rotational mass.

The rotational mechanical spring symbolic diagram is shown in Figure 2.32. The torque is given by

$$\frac{dQ}{dt} = K(\omega_2 - \omega_1) = K\omega_{21} \qquad (2.63)$$

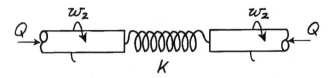

**Figure 2.32.
Rotational spring.**

or by

$$Q = K \int \omega_{21} dt = K\theta_{21} \qquad (2.64)$$

This equation is essentially identical to the equation for the pure translational spring. The capital letter K is used to distinguish rotational springs from translational springs.

What to Commit to Memory

Figure 2.33 shows the three fundamental rotational mechanical components, their symbolic diagrams and the equations you should always remember. Figure 2.34 provides a summary of the symbolic diagrams, operational block diagrams, and impedances for these building blocks. Go over each of these and note how similar they are to those shown in Figure 2.26 for the mechanical translational elements.

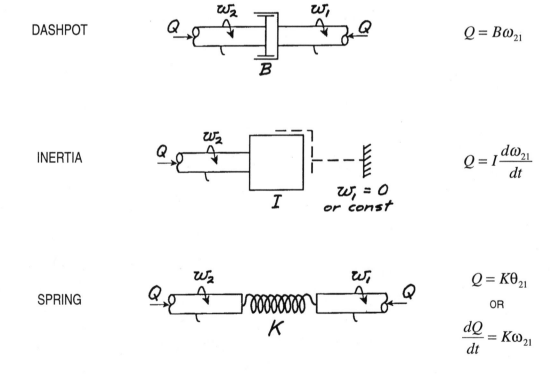

DASHPOT $\qquad Q = B\omega_{21}$

INERTIA $\qquad Q = I\dfrac{d\omega_{21}}{dt}$

SPRING $\qquad Q = K\theta_{21}$

OR

$\dfrac{dQ}{dt} = K\omega_{21}$

Figure 2.33. Rotational mechanical elements.

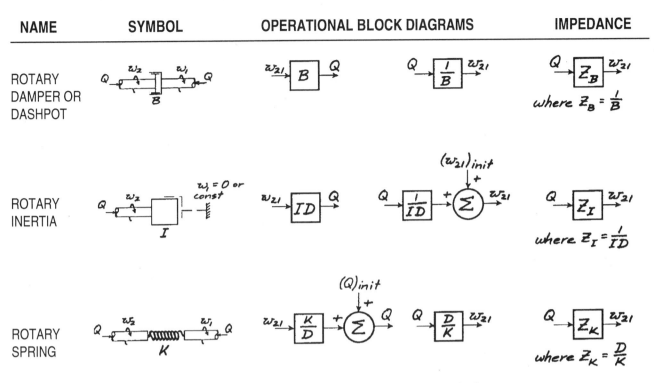

NAME	SYMBOL	OPERATIONAL BLOCK DIAGRAMS	IMPEDANCE

Figure 2.34. Summary of rotational mechanical elements.

2.5 Fluid Elements

Concepts of Pressure and Flow

A fluid is matter in a state which cannot sustain a shear force when it is at rest. For example, Figure 2.35 shows a circular "shear box" filled with a fluid. A minute force F will cause the contents of the upper half to spill. When the box is filled with a solid, a large force is required to shear the material. A fluid can be a liquid or a gas. The only difference between the two is related to the ease with which a gas can be compressed relative to a liquid. A fluid can be looked on as a quantity of small particles of matter that behave in such a manner that properties (pressure, mass per unit volume, etc.) can be measured at every point in the fluid.

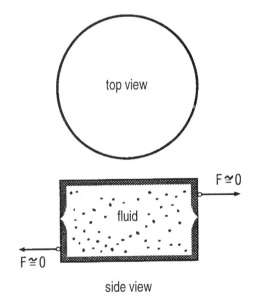

Figure 2.35. Circular box filled with a fluid. Upper half will slide over lower half immediately when a force F is applied.

Pressure in a fluid is defined as force per unit of area exerted in a direction that is perpendicular to the unit area. Pressure has the dimensions of force per length squared, the same as stress. Pressure is similar to the concept of normal stress in a solid.

In actuality, a fluid can sustain a shear force, but only when there is relative motion between fluid particles. For example, the flow of water in a river exerts a shear force on the bed of the river because the velocity of the water u varies as a function of the distance from the river bed y. At the river bed ($y = 0$) the water velocity is zero. Near the surface of the river ($y =$ depth of the water) the water velocity is near its maximum. A so-called *velocity gradient* therefore exists in the water causing relative motion between fluid particles. The velocity gradient can be expressed as a derivative (du/dy) where du is the incremental change in water velocity and dy is the incremental change in depth. The shear stress τ, defined as a force per unit area exerted tangentially to that unit area, is proportional to the velocity gradient. That is:

$$\tau = \mu \frac{du}{dy} \qquad (2.65)$$

where μ is the constant of proportionality, called the *absolute viscosity* of the fluid. Fluids which obey equation (2.65) are often called *Newtonian fluids*. (This, of course, is a simplification of fluid behavior; real fluids are more complex.)

Rate of flow of a fluid is defined as either the amount of mass or volume of fluid moving past a boundary in a unit of time. *Mass rate of flow* is generally used when dealing with a gas and *volume rate of flow* when dealing with a liquid.

The law of conservation of mass is often applied to a fluid to determine flow conditions at one point in a fluid, given the conditions at another. For example, Figure 2.36 shows two pipes with a gas flowing in one and a liquid in the other. At one end of each pipe the pressure is p_1 and the area is A_1. At the other the pressure is p_2 and the area is A_2. In the pipe carrying the gas, the mass density (mass per unit volume) varies and is ρ_1 at one end and ρ_2 at the other. In the pipe carrying the liquid, the mass density is a constant, ρ. The mass rate of flow q_{m1} entering the pipe must equal the mass rate of flow q_{m2} exiting the pipe since no mass is being stored inside the pipe. That is

$$q_{m1} = q_{m2}$$

or

$$\rho_1 A_1 U_1 = \rho_2 A_2 U_2 \qquad (2.66)$$

(This is commonly called the "mass continuity" equation.)

Note from equation (2.66) that the units of mass rate of flow are slugs/ft^3 $\times$ ft^2 $\times$ ft/sec = slugs/sec. (Note: a "slug" is the amount of mass which accelerates at the rate of 1 ft/sec^2 under the action of 1 lb of force.)

If the fluid is incompressible, ρ_1 and ρ_2 are equal to ρ which cancels out of both sides of equation (2.66). Equation (2.66) then reduces to

$$A_1 U_1 = A_2 U_2 \qquad (2.67)$$

That is, the volume rate of flow q_v must be equal at both ends of the pipe.

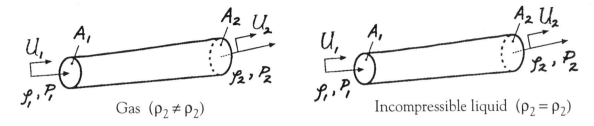

Gas ($\rho_2 \neq \rho_2$) Incompressible liquid ($\rho_2 = \rho_2$)

Figure 2.36. Two pipes with a gas flowing in one and a liquid in the other.

Another very useful equation can be found by applying Newton's laws of motion to an *ideal fluid* in *steady flow* conditions. An ideal fluid is one that is incompressible and has no viscosity. Steady flow means that particles of fluid are following a path in space that does not vary with time. Such paths are often called *streamlines* and bundles of streamlines are called *streamtubes*. A streamline or streamtube looks and behaves exactly like the tubes shown in Figure 2.36.

Let's examine the streamtube shown in Figure 2.37. If the pressure p_1, elevation h_1, and velocity U_1 of a point along the tube is known, then the pressure p_2, elevation h_2, and velocity U_2 of another point along the tube are related by

$$\left(p_2 - p_1\right) + \rho g\left(h_2 - h_1\right)$$

$$+ \frac{\rho}{2}\left(U_2^2 - U_1^2\right) = 0 \qquad (2.68)$$

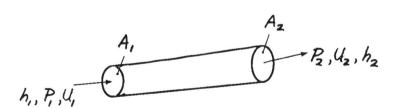

Figure 2.37. A streamtube.

Equation (2.68) is often called *Bernoulli's equation*. It is extremely useful when dealing with fluids and will be now used, together with the mass continuity equations given by equations (2.66) and (2.67), to describe equations for the three fundamental fluid elements.

As we know, power is the rate at which work is done. Referring again to the streamtube in Figure 2.37, we can write an expression for the power required to force the fluid into the entrance of the tube at point 1 as

$$Power_1 = \frac{dW_1}{dt} = p_1\frac{dvol}{dt} = p_1 q_v \qquad (2.69)$$

(See Appendix B.4 for more details.) Similarly, the power expended by the fluid exiting the streamtube can be found by

$$Power_2 = p_2 q_v \qquad (2.70)$$

Subtracting (2.70) from (2.69) gives the net power as the product of the pressure difference across the tube ends and the volume rate of flow through the tube. That is,

$$Power_1 - Power_2 = p_1 q_v - p_2 q_v$$

$$= p_{12} q_v \qquad (2.71)$$

The Fluid Resistor

Figure 2.38 shows a symbolic (circuit) diagram of a fluid resistor along with the fundamental describing equations for an *ideal fluid resistor*. Due to the form of Bernoulli's Equation, most fluid resistors are nonlinear. For example, sharp-edged orifices are frequently used in fluid systems to measure flow. Figure 2.39 shows such an orifice installed in a section of pipe. If the pipe is level so that gravity has no effect on the flow, then from equation (2.68) we can write

$$\left(p_2 - p_1\right) + \frac{\rho}{2}\left(U_2^2 - U_1^2\right) = 0 \qquad (2.72)$$

From equation (2.67) we can write

$$U_2 = \frac{A_1}{A_2} U_1 \qquad (2.73)$$

Substituting (2.73) into (2.72) gives

$$\left(p_2 - p_1\right) + \frac{\rho}{2}\left(\frac{A_1^2}{A_2^2}U_1^2 - U_1^2\right) = 0$$

or

$$\left(p_2 - p_1\right) + \frac{\rho}{2}U_1^2\left(\frac{A_1^2}{A_2^2} - 1\right) = 0 \qquad (2.74)$$

You can see that if A_1 is much smaller than A_2, then the quantity A_1^2 / A_2^2 will be much smaller than 1 and can be neglected. Equation (2.74) then reduces to

$$\left(p_2 - p_1\right) = \frac{\rho}{2}U_1^2 \qquad (2.75)$$

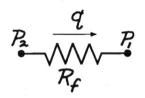

(a). Symbolic or circuit diagram.

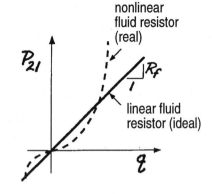

(b). Graphical representation.

$$q = \frac{P_2 - P_1}{R_f} = \frac{P_{21}}{R_f}$$

or

$$P_{21} = R_f q$$

(c). Describing equations for ideal fluid resistor.

Figure 2.38. The ideal fluid resistor

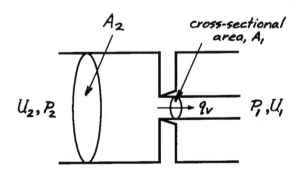

Figure 2.39. A sharp-edged orifice.

Using the relationship

$$q_v = A_1 U_1$$

we can also write (2.75) as

$$p_{21} = \left(\frac{\rho}{2A_1^2}\right) q_v^2$$

or more precisely as

$$p_{21} = \left(\frac{\rho}{2A_1^2}\right) q_v |q_v| \qquad (2.76)$$

where the absolute value allows the sign of p_{21} to change as the flow direction changes. You can see that the pressure difference is a function of the square of the volume rate of flow, causing the orifice to be a nonlinear resistor.

Of course equation (2.76) could be linearized about an operating point defined as q_{vo}, using Taylor's Theorem described in Appendix A, if a linear resistor were required. However, there *are* linear fluid resistors. Incompressible flow through a very small diameter tube (capillary) results in a linear relationship with the pressure differential across the ends of the tube and the volume rate of flow through the tube. Many soils and porous plugs made out of cinder material exhibit a linear relationship between pressure and flow.

A capillary tube is basically an orifice whose length is much greater than its diameter. The flow in such tubes is laminar; that is, it moves in layers, or laminas, and fluid particles do not bounce around from one layer to another. The

pressure differential across the ends of such a tube is related to the volume rate of flow through the tube by

$$p_{21} = \left(\frac{128\mu l}{\pi D^4}\right) q_v \qquad (2.77)$$

where D is the tube diameter, l is the tube length, and μ is the absolute viscosity of the liquid.

Compare (2.77) with the equation for an ideal electrical resistor

$$V_{21} = Ri$$

If we let $128\mu l / \pi D^4 = R_f$ in (2.77) then we can rewrite (2.77) as

$$\boxed{p_{21} = R_f q_v} \qquad (2.78)$$

A comparison of (2.78) with the electrical resistor equation shows they are completely analogous when the volume rate of flow is taken as the analog of electrical current and pressure differential is taken as the analog of electrical voltage differential.

The energy delivered to a fluid resistor in the interval from $t = t_a$ to $t = t_b$ is

$$E_b - E_a = \int_{t_a}^{t_b} p_{21} q_v dt \qquad (2.79)$$

Substituting equation (2.78) into (2.79) and eliminating q_v gives

$$E_b - E_a = \int_{t_a}^{t_b} p_{21}\left(\frac{p_{21}}{R_f}\right) dt = \frac{1}{R_f}\int_{t_a}^{t_b} p_{21}^2 dt \qquad (2.80)$$

Alternatively, we can use equation (2.78) to eliminate p_{21}

$$E_b - E_a = \int_{t_a}^{t_b} \left(R_f q_v \right) q_v dt$$

$$= R_f \int_{t_a}^{t_b} q_v^2 dt \qquad (2.81)$$

Just as we saw with the electrical resistor, equations (2.80) and (2.81) reveal that a fixed resistor dissipates power regardless of the direction of the fluid flow. *Mathematically, an ideal fluid resistor behaves exactly like an ideal electrical resistor when the voltage-pressure and current-flow analogy is used.*

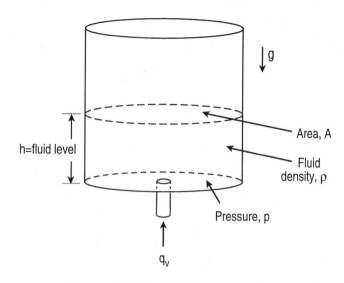

Figure 2.40. Fluid reservoir.

The Fluid Capacitor

Fluid capacitors are found in numerous hydraulic and pneumatic systems. Reservoirs, pressurized tanks, spring-loaded accumulators, and air-charged accumulators are examples of commonly encountered fluid capacitors.

An open reservoir is often used in a hydraulic system as a capacitor. Figure 2.40 shows the general arrangement of such a capacitor. A volume rate of flow q_v enters the bottom of the tank causing the level of the tank h to increase. This increased fluid level also increases the pressure p at the bottom of the tank.

The mass continuity equation applied to the tank gives us

$$q_v = A \frac{dh}{dt} \qquad (2.82)$$

We can also see that the pressure at the bottom of the tank is equal to the weight of the fluid in the tank divided by the area. That is

$$p = \frac{\rho g h A}{A} = \rho g h \qquad (2.83)$$

Equation (2.83) can be rearranged and then differentiated to give

$$\frac{dh}{dt} = \frac{1}{\rho g} \frac{dp}{dt} \qquad (2.84)$$

Substituting (2.84) into (2.82) to eliminate dh/dt gives

$$q_v = \left(\frac{A}{\rho g} \right) \frac{dp}{dt} \qquad (2.85)$$

Compare (2.85) to the equation for an ideal electrical capacitor:

$$i = C \frac{dV_{21}}{dt}$$

Clearly the two equations are mathematically the same if we define $(A/\rho g)$ to be a fluid capacitance C_f and we use the electrical voltage–fluid pressure and electrical current–fluid flow analogy. Equation (2.85) can then be written as

$$q_v = C_f \frac{dp}{dt} \qquad (2.86)$$

where $C_f = A/\rho g$.

An accumulator is another form of fluid capacitor. A spring-loaded accumulator is shown in Figure 2.41. In this type of accumulator a spring rather than gravity provides the pressure increase. A volume rate of flow q_v entering the bottom of the tank causes the spring to compress a distance x. This increases the pressure p in the tank.

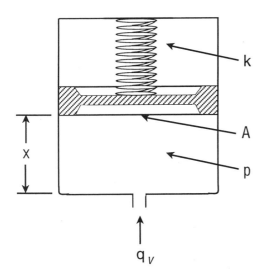

Figure 2.41. Spring-loaded accumulator.

The mass continuity equation applied to the tank gives

$$q_v = A \frac{dx}{dt} \qquad (2.87)$$

The pressure in the tank is also equal to the force exerted by the spring on the fluid divided by the area. That is

$$p = \frac{kx}{A} \qquad (2.88)$$

where k is the spring constant. Equation (2.88) can be rearranged and then differentiated to give

$$\frac{dx}{dt} = \frac{A}{k} \frac{dp}{dt} \qquad (2.89)$$

Substituting (2.89) into (2.87) to eliminate dx/dt gives

$$q_v = \left(\frac{A^2}{k}\right) \frac{dp}{dt} \qquad (2.90)$$

Equation (2.90) can then be written as a fluid capacitor

$$q_v = C_f \frac{dp}{dt} \qquad (2.91)$$

where $C_f = A^2/k$.

The energy delivered to a fluid capacitor in the time interval from $t = t_a$ to $t = t_b$ is given by

$$E_b - E_a = \int_{t_a}^{t_b} p_{12} q_v dt \qquad (2.92)$$

Equation (2.91) can be rewritten using differentials as

$$q_v dt = C_f dp \qquad \begin{matrix}(2.91)\\ \text{rewritten}\end{matrix}$$

Substituting $C_f dp$ for $q_v dt$ in equation (2.92) gives

$$E_b - E_a = C_f \int_{p_a}^{p_b} p \, dp \qquad (2.93)$$

Equation (2.93) can be integrated to give

$$E_b - E_a = C_f \left[\frac{p^2}{2} \right]_{p_a}^{p_b}$$

$$= C_f \frac{p_b^2}{2} - C_f \frac{p_a^2}{2} \qquad (2.94)$$

The quantity $C_f(p_a^2 / 2)$ is the energy that was initially stored in the capacitor at $t = t_a$ and $C_f(p_b^2 / 2)$ represents the energy stored at time $t = t_b$. During the time interval $t_b - t_a$, the energy $E_b - E_a$ was added to the fluid capacitor. You can see that fluid capacitors, like their electrical cousins, can store energy. This energy can be retrieved and used later when needed.

The Fluid Inductor

A mass of fluid in motion is quite similar to a solid mass in motion. The fluid mass has inertia and a force is required to accelerate or decelerate the fluid. Figure 2.42 shows an *ideal* (no viscosity and hence no friction forces) *incompressible* fluid in *unsteady* (flow velocity is not a constant) flow through a pipe. Let's apply Newton's laws of motion to the mass of fluid in the pipe.

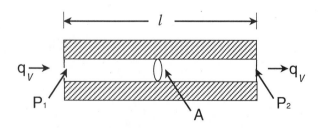

Figure 2.42. Ideal incompressible fluid in unsteady flow through a pipe.

The fluid mass is $\rho A l$, the net force acting on this mass is $A(p_1 - p_2)$, and the acceleration of the fluid mass is dU / dt. From Newton's laws we can write

$$A(p_1 - p_2) = \rho A l \frac{dU}{dt} \qquad (2.95)$$

Since $q_v = AU$ and $p_{12} = p_1 - p_2$, equation (2.95) can also be rewritten as

$$p_{12} = \left(\frac{\rho l}{A} \right) \frac{dq_v}{dt} \qquad (2.96)$$

Compare equation (2.96) with the equation for an electrical inductor:

$$V_{21} = L \frac{di}{dt}$$

Clearly this equation is analogous to (2.96) if we define $(\rho l / A)$ to be a fluid inductance I_f and we use the electrical voltage–fluid pressure and electrical current–fluid flow analogy. Equation (2.96) can therefore be written as

$$\boxed{p_{12} = I_f \frac{dq_v}{dt}} \qquad (2.97)$$

You can see that, mathematically, *an ideal fluid inductor behaves exactly like an ideal electrical inductor.*

The energy delivered to a fluid inductor in the time interval from $t = t_a$ to $t = t_b$ is given by

$$E_b - E_a = \int_{t_a}^{t_b} p_{12} q_v \, dt$$

Substituting equation (2.97) into this equation to eliminate p_{12} gives

$$E_b - E_a = \int_{t_a}^{t_b} \left(I_f \frac{dq_v}{dt} \right) q_v \, dt$$

$$= I_f \int_{q_{va}}^{q_{vb}} q_v \, dq_v \qquad (2.98)$$

Equation (2.98) can be integrated to give

$$E_b - E_a = I_f \left[\frac{q_v^2}{2} \right]_{q_{va}}^{q_{vb}}$$

$$= I_f \frac{q_{vb}^2}{2} - I_f \frac{q_{va}^2}{2} \qquad (2.99)$$

The quantity $I_f(q_{va}^2 / 2)$ is the kinetic energy associated with the initial flow rate at $t = t_a$ and $I_f(q_{vb}^2 / 2)$ is kinetic energy associated with the final flow rate at $t = t_b$. During the time interval $t_b - t_a$, the kinetic energy $E_b - E_a$ was added to the fluid.

What to Commit to Memory

You should always remember that equations associated with fluid mechanics are based on applying the mass continuity (conservation of mass) and Newton's motion laws to a fluid. You should be able to write mass continuity from memory in the form

$$\rho_1 A_1 U_1 = \rho_2 A_2 U_2$$

You should also be able to write Bernoulli's equation from memory and know that it strictly applies to a frictionless fluid in steady flow. Remember the equation in one of these two forms

$$\left(p_2 - p_1 \right) + \rho g \left(h_2 - h_1 \right) + \frac{\rho}{2} \left(U_2^2 - U_1^2 \right) = 0$$

or

$$p_2 + \rho g h_2 + \frac{\rho}{2} U_2^2 = p_1 + \rho g h_1 + \frac{\rho}{2} U_1^2$$

If you remember these equations, you will always be able to derive the elemental equations for a fluid resistor, fluid capacitor, and a fluid inductor. Also keep in mind when working with fluid systems that the three fundamental building blocks are not as easy to spot as are their electrical counterparts. So make it a habit to try to identify the three fluid building blocks whenever you encounter a fluid system. (You could start with the fresh and hot water supplies in your home.)

Figure 2.43 shows the symbolic diagrams, operational block diagrams, and impedances for the three fundamental fluid elements. As you review these, note how similar they are to the three fundamental electrical and mechanical elements in Figures 2.15, 2.26, and 2.34.

2.6 Thermal Elements
Concepts of Temperature and Heat

The concepts of *temperature* and *heat* are encountered in almost all engineering systems. Heat must be dissipated from electrical circuits or the elements will burn out. Heat produced by friction between mechanical elements can cause them to seize if the heat is not removed. Fluids are frequently used to transfer heat from one location to another.

We qualitatively think of temperature as a measure of how "hot" or "cold" an object is. This implies that temperature is a relative term, and indeed it is. A quantitative measure of tempera-

NAME	SYMBOL	OPERATIONAL BLOCK DIAGRAMS	IMPEDANCE

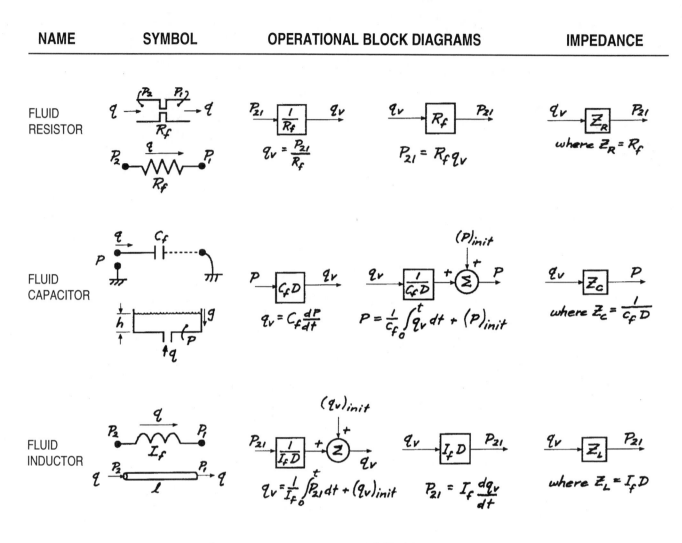

Figure 2.43. Summary of fluid elements.

ture is obtained with a thermometer and a variety of scales are in use today. For example, the Celsius or centigrade scale was developed based on the freezing and boiling points of water. The temperature at which solid water (ice) changes to liquid water is equal to 0 degrees Celsius (0°C) and the temperature at which liquid water changes to gaseous water (steam) is equal to 100 degrees Celsius (100°C).

When two bodies at different temperatures are brought in contact with one another, *heat* flows from the body with the higher temperature to the body with the lower temperature. When this happens, energy is transferred from the hotter to the colder body. Heat is defined as the energy that is transferred from one body or system to another as the result of the temperature difference between the two.

Concepts of Work, Power, and Energy in Thermal Elements

When mechanical work is done on a body, its temperature will rise unless heat is removed from the body. The First Law of Thermodynamics states that work and heat can be converted from one to the other. Consequently, the units of heat and work are equivalent. In the English system, temperature is usually measured in Fahrenheit degrees and heat in British Thermal Units (BTU). One BTU is equivalent to 778.172 ft-lbs. In the SI system, temperature is measured in degrees centigrade and heat in joules. You will recall from our study of electrical elements that one joule is equal to one watt-sec, or 0.737 ft-lbs.

There are three basic ways in which heat can be transferred from a hotter to a colder body: *convection*, *radiation* and *conduction*. Convection-type heat transfer takes place primarily in liquids and gases. Heat is transferred as the result of matter moving from one location to another due to currents set up by the temperature differentials. Radiation-type heat transfer takes place as a result of energy carried by electromagnetic waves. Conduction-type heat transfer usually involves substances in the solid phase. Heat is transferred at the atomic level without any visible motion of matter.

Thermal Resistance

All material offers some resistance to heat flow. When a material offers a large degree of resistance it is often called an insulator. Figure 2.44 shows a section of an insulative material.

It has been found through experimentation

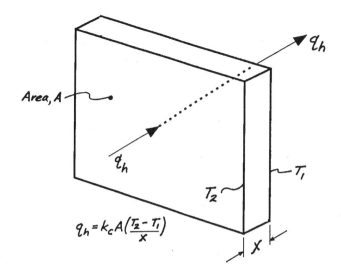

Figure 2.44. Heat flow through a conducting material.

that the rate of heat transfer q_h between two surfaces of area A is proportional to the area and the temperature gradient dT/dx. In equation form,

$$q_h \propto -A\frac{dT}{dx} \qquad (2.100)$$

By using a constant of proportionality k_c we can write equation (2.100) as

$$q_h = -k_c A\frac{dT}{dx} \qquad (2.101)$$

The constant of proportionality k_c is called the thermal conductivity. The negative sign indicates that heat flows from the direction of decreasing temperature.

Applying equation (2.101) to Figure 2.44 gives

$$q_h = k_c A\frac{(T_2 - T_1)}{X} \qquad (2.102)$$

Symbol	Meaning	English units	SI units
q_h	rate of heat flow	BTU/hr	joules/sec
k_c	thermal conductivity	$\dfrac{\text{BTU}}{\text{hr - ft}^2 \text{ - (deg F / ft)}}$	$\dfrac{\text{joules}}{\text{sec - m}^2 \text{ - (deg C / m)}}$
A	cross-sectional area	ft^2	m^2
T_2	hotter temperature	deg F	deg C
T_1	colder temperature	deg F	deg C
X	thickness of material	ft	m

Take careful note of the units in this equation. They are given above in both English and SI units.

If we rewrite equation (2.102) in the form

$$T_2 - T_1 = T_{21} = \left(\frac{X}{k_c A}\right) q_h \qquad (2.103)$$

and then compare this with the equation for an electrical resistor

$$V_{21} = Ri$$

we can see that the two equations are identical provided we think of $X/k_c A$ as a resistor (thermal), temperature as a voltage, and rate of heat flow as an electrical current. Drawing this analogy, we can then write (2.103) as

$$\boxed{T_{21} = R_t q_h} \qquad (2.104)$$

where

$$R_t = \frac{X}{k_c A}$$

We must be very careful with this analogy when it comes to looking at power relationships. The rate of heat flow q_h corresponds to power, as a consequence of the equivalency of heat and work stated by the First Law of Thermodynamics. Consequently, power is given by equation (2.102), or in terms of R_t as

$$Power = q_h = \frac{1}{R_t} T_{21} \qquad (2.105)$$

Thermal Capacitance

All materials have some capacity to store heat. The capacity of a material to store heat is called its *specific heat*. It is defined as the amount of heat per unit mass that must be added to the material to raise its temperature one degree. Specific heat c is, in general, a function of temperature and can be written as

$$c = c(T) = \frac{1}{m}\frac{dQ}{dT} \qquad (2.106)$$

where m = mass of the material

dQ = amount of heat required

dT = temperature rise due to addition of dQ

If we restrict attention to small variations of temperature, then c can be taken as a constant. In that case equation (2.106) can be written as

$$dQ = cmdT \qquad (2.107)$$

Dividing through by an incremental time dt, gives

$$\frac{dQ}{dt} = q_h = cm\frac{dT}{dt} \qquad (2.108)$$

Let's compare equation (2.108) with the equation for an electrical capacitor:

$$i = C\frac{dV}{dt}$$

You can see that the two equations are identical if we think of cm as a capacitor (thermal), temperature as a voltage, and rate of heat flow as an electrical current. In this case we rewrite equation (2.108) as

$$q_h = C_t\frac{dT}{dt} \qquad (2.109)$$

where

$$C_t = cm$$

Once again, you must be careful with this analogy when it comes to looking at power relationships. Quite often we are interested in how much heat is stored in a material. This can be obtained by integrating equation (2.109). That is,

$$Q = m\int_{T_1}^{T_2} cdT \qquad (2.110)$$

If c is taken as a constant, then equation (2.110) can be integrated giving

$$Q = mc(T_2 - T_1) \qquad (2.111)$$

Thermal Inductance

There is no known thermal phenomena which stores thermal energy as a function of the rate of change of the heat flow rate. Consequently no analogies with electrical inductance can be drawn.

What to Commit to Memory

The equations for thermal resistance and thermal capacitance given above will likely serve many of your needs when modeling engineering systems. You should commit to memory the analogies with the electrical resistor and capacitor.

Figure 2.45 shows symbolic diagrams, operational diagrams and impedances for the two fundamental thermal elements. Compare these with the electrical, mechanical, and fluid "resistors" and "capacitors." You will see that the equations are identical. However, don't forget that power in thermal systems is *not* equal to the product of the across variable (temperature) and the through variable (heat flow rate), as is the case with the electrical, mechanical, and fluid systems. The through variable for thermal systems, heat flow rate, is itself power.

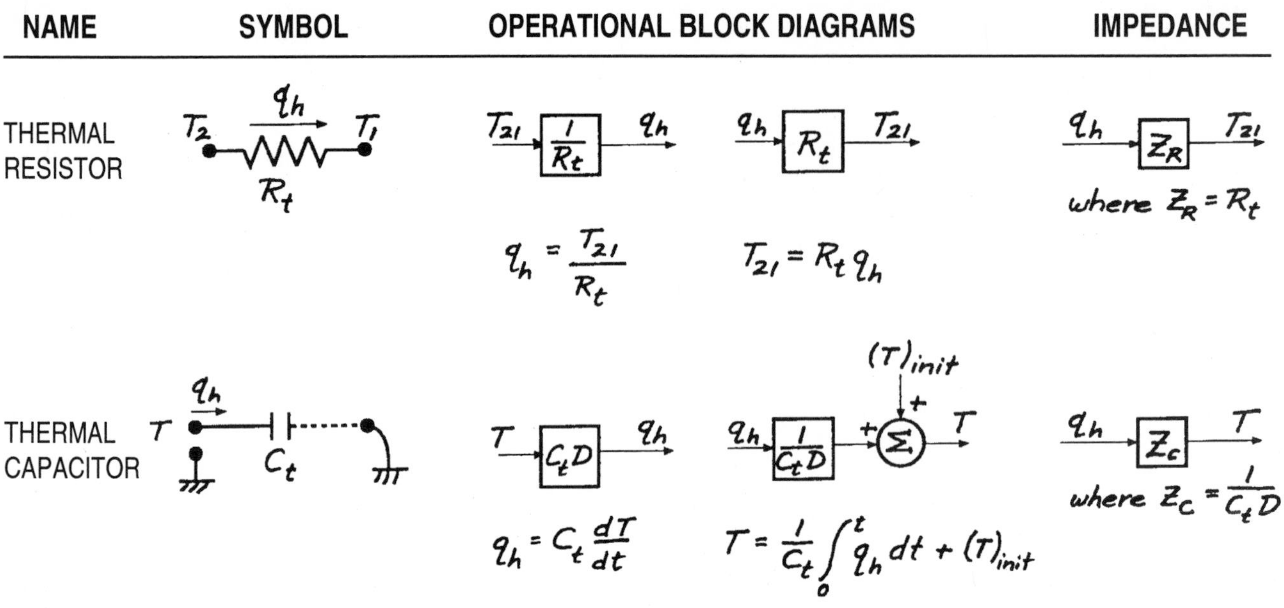

NAME	SYMBOL	OPERATIONAL BLOCK DIAGRAMS		IMPEDANCE

THERMAL RESISTOR

$$q_h = \frac{T_{21}}{R_t} \qquad T_{21} = R_t q_h$$

where $Z_R = R_t$

THERMAL CAPACITOR

$$q_h = C_t \frac{dT}{dt} \qquad T = \frac{1}{C_t}\int_0^t q_h\, dt + (T)_{init}$$

where $Z_C = \frac{1}{C_t D}$

Figure 2.45. Summary of thermal elements.

2.7 The Importance of Analogies

This chapter is probably the most important chapter in this book. It covers what I believe to be the most important fundamentals of engineering. If you understand this material and commit to memory the basic formulas, you will have a solid grounding for any engineering work.

Throughout this chapter I have repeatedly pointed out how the equations describing the three fundamental components in electrical, mechanical, and fluid systems are identical if the proper analogies are drawn between *across* and *through* variables. I have also pointed out that the *product* of the across and through variables for these three vastly different engineering system building blocks is equal to power. Table 2.3 summarizes what we've covered so far.

Note that I have included thermal systems in this summary, but placed them in a box to remind you that the product of temperature (across variable) and rate of heat flow (through variable) is not power.

The real beauty of analogies lies in the way in which mathematics unifies these diverse fields of engineering into one subject. That means tools developed for solving problems in one field can be used to solve problems in another. This is an important concept, since some fields, particularly electrical engineering, have rich sets of problem-solving tools that are fully applicable to other engineering fields. The more you work with engineering problems and these analogies, the more intuitive your engineering skills will become. The result—the good engineering judgment so essential to success in any type of engineering field.

Table 2.3. Analogies summary.

System	Across Variable	Through Variable	Power Equation	Resistor Equation	Capacitor Equation	Inductor Equation
ELECTRICAL	Voltage (V)	Current (i)	$V \times i$	$V = Ri$	$i = C\dfrac{dV}{dt}$	$V = L\dfrac{di}{dt}$
MECHANICAL TRANSLATION	Velocity (v)	Force (F)	$v \times F$	$v = \dfrac{1}{b}F$	$F = m\dfrac{dv}{dt}$	$v = \dfrac{1}{k}\dfrac{dF}{dt}$
MECHANICAL ROTATIONAL	Angular Velocity (ω)	Torque (Q)	$\omega \times Q$	$\omega = \dfrac{1}{B}Q$	$Q = I\dfrac{d\omega}{dt}$	$\omega = \dfrac{1}{K}\dfrac{dQ}{dt}$
FLUID	Pressure (p)	Flow Rate (q_v)	$p \times q_v$	$p = R_f q_v$	$q_v = C_f\dfrac{dp}{dt}$	$p = I_f\dfrac{dq_v}{dt}$
THERMAL	Temperature (T)	Heat Flow Rate (q_h)	q_h	$T = R_t q_h$	$q_h = C_t\dfrac{dT}{dt}$	None

Chapter

3

Constructing First-Order Math Models

Objectives

When you have completed this chapter, you will be able to:

■ **Develop simple math models of engineering systems.**

■ **Define the *path-vertex-elemental equation,* the *impedance,* the *operational block diagram,* and the *free-body diagram* methods for developing math models and know how to use each method.**

■ **Recognize when the fundamental elements are connected in *series* or in *parallel* and use the associated simplifying equations in the development of math models.**

■ **Define and understand what is meant when a math model is described as a *first-* or *second-order ordinary linear differential equation with constant coefficients.*

■ **Recognize and use the power of the *impedance method* for rapidly developing math models.**

3.1 Introduction to Math Models

In the previous chapter you learned how to develop mathematical models of single-element, ideal components found in electrical, mechanical, fluid, and thermal systems. You can model numerous systems found in engineering by breaking the system down into these fundamental elements and then combining the elemental equations to form a mathematical model of the complete system. An electrical system, for example, might be modeled as a combination of an ideal resistor and an ideal capacitor. Or a mechanical system might be modeled as a combination of an ideal spring and an ideal mass.

In this chapter, you will learn that when equations for the ideal elements are combined, a mathematical model referred to as a *differential equation* is obtained. This equation describes the behavior in time of the *output* (or *response*) variable of the system to a time-varying *input* or *forcing* variable. You will also learn that the coefficients of the differential equation are related to the constants associated with the ideal elemental equations (that is, the value of the resistor, capacitor, inductor, mass, spring, etc.). The ideal element constants are combined and form *parameters* which govern the most basic characteristics of the differential equation.

Based on my own experience and observation of other engineers, I've discovered that it is generally much harder to create mathematical models of systems (that is, derive the governing differential equations) than it is to solve the equations once they have been derived. In fact, there are many excellent software programs available now that can be run on an inexpensive PC. These programs can be used to prepare math models of engineering systems and to solve the resultant equations. But these programs can be dangerous! You *must* have the engineering basics before you can model an engineering system, and you need to know how to solve at least basic equations so you can judge whether or not the output you are getting from a computer program makes sense.

I consider the material covered in this chapter to be the true dividing line between a "good" engineer and a mediocre one. A good engineer has internalized this material so that he or she is able to conceptualize an engineering system—to "see" the differential equations that describe its behavior—almost automatically. When you step on the accelerator of your car, for example, you should be able to visualize the graphs of thrust vs. velocity and resistance vs. velocity that determine the acceleration of its mass. Don't expect this to happen overnight, however. It comes with lots of practice!

I spend a great deal of time in this chapter explaining techniques and developing tools that you can use to make the equation derivation process easier. More emphasis will be placed on the methods that I have found produce answers in the shortest period of time and that are easiest to remember. To keep things as simple as possible, only engineering systems that contain two fundamental elements will be discussed in this chapter. While this may at first seem trivial, you will discover that nearly every engineering system contains at least one component that can be described fairly accurately by a combination of two ideal elements.

3.2 Tools for Developing Math Models

Four methods for developing mathematical models of engineering systems are presented in this section. They are usually referred to by the following names:

- *Path-Vertex-Elemental Equation Method*
- *Impedance Method*
- *Operational Block Diagram Method*
- *Free-body Diagram Method*

The path-vertex-elemental equation method has a variety of names depending on the field of engineering in which you're involved. In electrical engineering the method is referred to as Kirchoff's Laws (voltage and current). Other fields often call this method "derivation by first principles."

The impedance method and the operational block diagram method seem to be used mostly by electrical engineers. That's too bad, because both methods are extremely powerful when applied to other fields as well. Learn them and you will save time and gain new insights into the systems you are trying to design or analyze, no matter what type of system it is.

The free-body diagram is most commonly used in mechanical engineering. It is an extremely powerful method and, since nearly all engineering systems have some mechanical components, all engineers should learn how to use it.

Again, I will present the first three methods using simple electrical circuits. This will allow you to better understand the important concepts of series and parallel "circuits" and how these concepts can be used to simplify the development of math models in general.

Path-Vertex-Elemental Equation Method

When several of the fundamental elements described in the previous chapter are connected together, the result looks like a network of highways with junctions and interconnections. For example, Figure 3.1 shows a complex electrical circuit comprised of interconnected resistors, capacitors, and inductors. The point where two or more elemental components are connected is called the *vertex*. The highway leading from one vertex to another is called the *path*.

A vertex is always associated with the elemental *across* variables. At each vertex, the across variables for each interconnected element are identical.

A path is always associated with the elemental *through* variables. The sum of the through variables flowing into or out of a vertex must equal zero. For example, in Figure 3.1, the currents flowing into and out of the vertex labelled V_2 must equal zero. That is, i_1 (flowing into the vertex) must equal the sum of i_2 and i_3 (flowing out of the vertex).

The path-vertex equations are often called "laws" and are frequently named after the person who discovered them. In electrical engineering, the path-vertex equations are called Kirchoff's voltage and current laws. While it's great to honor past engineers for their contributions, don't let names confuse you. You learned in Chapter 2 that you can create a "circuit" for any

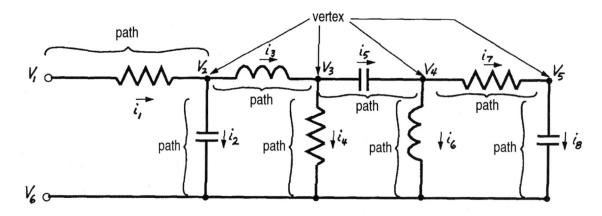

Figure 3.1. A complex electrical circuit showing *vertices* and *paths*. Note that V_1, V_2, ..., V_5 are "across" variables and i_1, i_2, ... i_8 are "through" variables.

electrical, mechanical, fluid, and thermal element. The path-vertex-elemental equation method is simply a procedure that allows you to derive equations for any "circuit," whether it is electrical, mechanical, fluid, or thermal, by writing the elemental equations associated with each element in the "circuit" and using the path and vertex laws associated with the across and through variables at each vertex. Now let's look at this method in more detail.

Series and Parallel Circuits

When two fundamental engineering system elements are connected together, they can only be connected in one of two ways: in series or in parallel. Figure 3.2a shows two electrical resistors (they could just as well be mechanical, fluid, or thermal resistors) connected in series and Figure 3.2b shows them connected in parallel. The voltages (across variable) at each of the vertices are labelled and the electrical current flowing (through variable) in each resistor and

into each vertex is shown. The directions of the currents are shown by arrows, but this is arbitrary since the relative values of the voltages are not given. In Figure 3.2a, there can only be one current and it flows through both resistors. In Figure 3.2b, the current flowing in the circuit splits and part flows through each resistor.

You can find the voltage V_1 at the juncture of the two resistors in Figure 3.2a by writing the elemental equation for the current flowing in each resistor. For R_a, the current is

$$i = \frac{V_2 - V_1}{R_a} \tag{3.1}$$

and for R_b, the current is

$$i = \frac{V_1 - V_0}{R_b} \tag{3.2}$$

Since the current flowing out of resistor R_a must flow into R_b, we can equate (3.1) and (3.2), giving

$$\frac{V_2 - V_1}{R_a} = \frac{V_1 - V_0}{R_b} \qquad (3.3)$$

Solving this equation for V_1 gives

$$V_1 = \frac{R_b}{R_a + R_b} V_2 - \frac{R_a}{R_a + R_b} V_0 \qquad (3.4)$$

If V_0 is taken as ground, or reference, and set to zero, then (3.4) becomes

$$V_1 = \frac{R_b}{R_a + R_b} V_2 \qquad (3.5)$$

Equation (3.5) indicates that the voltage V_1 is equal to the voltage V_2 times the ratio of the resistor R_a and the sum of the two resistors. If the two resistors were equal, the ratio would be

$$\frac{R_a}{R_a + R_b} = \frac{R}{R + R} = \frac{R}{2R} = \frac{1}{2}$$

If R_a were twice the size of R_b, the ratio would be 2/3. In essence, the circuit shown in Figure 3.2a is acting as a voltage divider (or constant multiplier). If V_2 is considered an input to the circuit and V_1 the output, then the circuit multiplies the input by the ratio $R_a / (R_a + R_b)$ and provides the product as the output.

The circuit given in Figure 3.2a and the describing equation given in (3.5) are extremely important. You will encounter them often throughout your career and they will be used many times throughout this book.

Now look at the circuit in Figure 3.2b. Note that the voltage across each resistor is the same and is equal to $V_1 - V_0$. The current i_a is given by

$$i_a = \frac{V_1 - V_0}{R_a} = \frac{V_1}{R_a} \qquad (3.6)$$

when V_0 is taken as the reference or zero voltage. Similarly the current i_b is given by

$$i_b = \frac{V_1 - V_0}{R_b} = \frac{V_1}{R_b} \qquad (3.7)$$

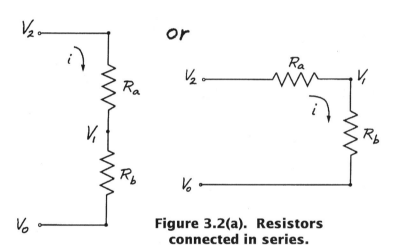

Figure 3.2(a). Resistors connected in series.

Figure 3.2(b). Resistors connected in parallel.

The current, i, flowing into the junction can be written as

$$i = i_a + i_b \tag{3.8}$$

since the sum of the currents flowing into a vertex must equal the sum of the currents flowing out of the junction.

Substituting (3.6) and (3.7) into (3.8) gives

$$i = \frac{V_1}{R_a} + \frac{V_1}{R_b} = \left(\frac{R_a + R_b}{R_a R_b} \right) V_1 \tag{3.9}$$

or

$$V_1 = \left(\frac{R_a R_b}{R_a + R_b} \right) i \tag{3.10}$$

Equation (3.10) reveals that resistors in parallel can be combined into an equivalent single resistor, R, given by

$$R = \frac{R_a R_b}{R_a + R_b} \tag{3.11}$$

as shown in Figure 3.3. You should note that, in essence, equation (3.11) allows you to reduce circuit complexity by replacing resistors in parallel with a single resistor.

Series and Parallel Circuits Involving One Energy Storage and One Energy Dissipative Element

So far we have only looked at resistors connected in series and in parallel. Figure 3.4a shows an electrical resistor and capacitor in series and Figure 3.4b shows the same elements connected in parallel.

Let's follow the same procedures used above and determine if we get similar equations. We will first write the elemental equations for the series circuit. The current through the resistor is

$$i = \frac{V_2 - V_1}{R} \tag{3.12}$$

The current through the capacitor is

$$i = C \frac{dV_1}{dt} \tag{3.13}$$

As before, let's equate (3.12) and (3.13) to eliminate i. That gives

$$\frac{V_2 - V_1}{R} = C \frac{dV_1}{dt} \tag{3.14}$$

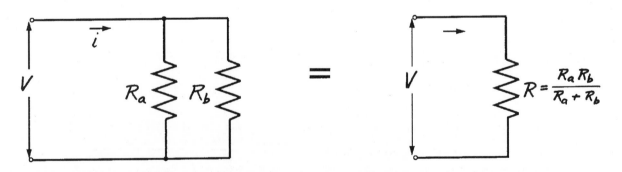

Figure 3.3. Reducing circuit complexity by combining elements in parallel.

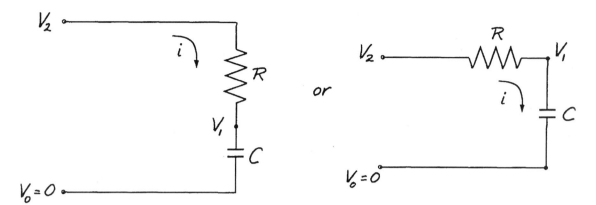

Figure 3.4(a). A resistor and capacitor in series.

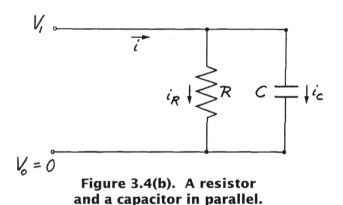

Figure 3.4(b). A resistor and a capacitor in parallel.

tion of the input voltage V_2, we did obtain a relationship between V_1 and V_2 that contains the time derivative of V_1, V_1 by itself, and V_2. Now let's check the units of each term in the equation to make sure they are correct. Each term on the left side of equation 3.15b must have the units of volts because only volts appear on the right side of the equation. That is,

$$\underbrace{RC\frac{dV_1}{dt}}_{\text{volts}} + \underbrace{V_1}_{\text{volts}} = \underbrace{V_2}_{\text{volts}}$$

Look at the first term. Since dV_1/dt has units of volts per unit of time, RC must have units of time. That is,

$$\left[RC\frac{dV_1}{dt} \right] = \sec \times \frac{\text{volts}}{\sec}$$

The product RC is called the *time constant* of the circuit and is given the Greek symbol, τ.

The relationship given by (3.15) is a *first-order linear ordinary differential equation with constant coefficients*. It is called a *differential equation* because it has both a variable and its derivative

Now let's rearrange this equation so all the terms containing V_1 are on the left side of the equal sign and all others are on the right side. That is

$$C\frac{dV_1}{dt} + \frac{1}{R}V_1 = \frac{1}{R}V_2 \qquad (3.15a)$$

Multiplying both sides by R gives

$$RC\frac{dV_1}{dt} + V_1 = V_2 \qquad (3.15b)$$

While we did not derive an equation which gives the output voltage, V_1, directly as a func-

in the same equation. It's called *ordinary* because it contains no partial derivative. It's called *first-order* because it involves only the first derivative of the variable. It is qualified as having *constant coefficients* [C and $1/R$ in (3.15a) and RC and 1 in (3.15b)] to distinguish it from similar equations that have coefficients which vary with time. It is called *linear* for reasons which will become clearer later.

You will find out later that (3.15) can be solved for V_1 (often called the *dependent*, *response*, or *output* variable) but the solution will give V_1 as a function of time (often called the *independent* variable) as well as a function of V_2 (often called the *input* or *forcing* variable).

Now let's derive the equation for the circuit in Figure 3.4b. The current flowing through the resistor is

$$i_R = \frac{V_1}{R} \qquad (3.16)$$

and through the capacitor is

$$i_C = C\frac{dV_1}{dt} \qquad (3.17)$$

The current summation at the junction gives

$$i = i_C + i_R \qquad (3.18)$$

Substituting (3.16) and (3.17) into (3.18) gives

$$i = C\frac{dV_1}{dt} + \frac{V_1}{R} \qquad (3.19)$$

Once again our final result is a first-order linear differential equation with constant coefficients. The equation is very similar to (3.15) in

that it involves the time derivative of V_1 and V_1 by itself.

In equation (3.19) I have purposely placed the time derivative of V_1 and V_1 on the right side of the equal sign to make a point. Ordinarily, you place the output (dependent) variable on the left side of an equation and the input (forcing) variable on the right. But in this case, which is which? Is V_1 the input or is i? If V_1 is the input as (3.19) implies, then we have succeeded in arriving at the desired equation and need go no further in solving the equation. That is, if V_1 is the input variable, then it is known as a function of time and therefore its first derivative can be determined. On the other hand, if i is the input variable then (3.19) should be rearranged in the form

$$RC\frac{dV_1}{dt} + V_1 = Ri \qquad (3.20)$$

Now compare (3.20) and (3.15b). Clearly they are identical except for the right sides. As I indicated with (3.15), you can solve (3.20) for V_1, but the solution will be a function of time as well as the input current, i.

Impedance Method

You discovered in the previous section that symbolic (circuit) diagrams can often be simplified for purposes of analysis by recognizing elements that are in series and parallel. You found out that resistors in series act as voltage dividers and input/output relations can be written directly without the need for lengthy equation derivations. You also discovered that resistors in parallel can be combined into an equivalent

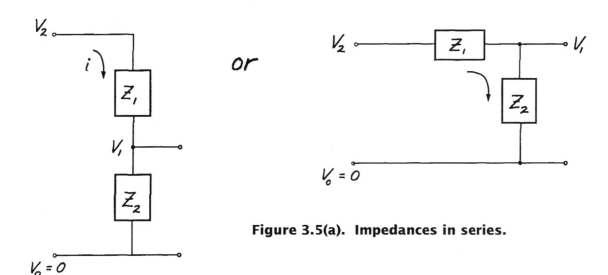

Figure 3.5(a). Impedances in series.

resistance. This too provides direct input/output relations without lengthy derivations.

In the last chapter you were introduced to the concept of impedance. You learned that inductors and capacitors can be represented as an electrical impedance just as a resistor can. You will now discover that any circuit element in series or in parallel can be treated just like a resistor when impedances are used. By combining the concepts of impedance and the tools for analyzing elements in series and in parallel, an extremely powerful method of deriving equations for complex circuits evolves.

Figure 3.5 shows the same two circuits shown in Figure 3.4 and studied using the path-vertex-elemental equation approach to deriving equations. All that has been changed between Figures 3.4 and 3.5 is the circuit element representation. The elements are shown in Figure 3.5 as generalized impedances.

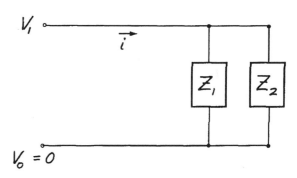

Figure 3.5(b). Impedances in parallel.

Using equation (3.5) as a guide, we can write the output voltage in terms of the input voltage and the ratio of the impedances:

$$V_1 = \frac{Z_2}{Z_1 + Z_2} V_2 \qquad (3.21)$$

where

$$Z_1 = R \qquad (3.22)$$

and

$$Z_2 = \frac{1}{CD} \qquad (3.23)$$

Substituting (3.22) and (3.23) into (3.21) gives

$$V_1 = \frac{\dfrac{1}{CD}}{R + \dfrac{1}{CD}} V_2 \qquad (3.24)$$

This equation can be simplified by treating the operator $1/D$ as if it were just an algebraic variable:

$$V_1 = \frac{\dfrac{1}{CD}}{\dfrac{RCD + 1}{CD}} V_2 = \frac{1}{RCD + 1} V_2 \qquad (3.25)$$

Cross-multiplying by $(RCD + 1)$ gives:

$$(RCD + 1)V_1 = RCDV_1 + V_1 = V_2 \qquad (3.26)$$

Now note that the operator D, or $d(\ )/dt$, is operating on V_1. Equation (3.26) can then be written as

$$RC\frac{dV_1}{dt} + V_1 = V_2 \qquad (3.27)$$

Comparing (3.27) to (3.15b) reveals they are identical.

Let's look at what we accomplished. Using simple algebra, we have taken the circuit shown in Figure 3.4a and derived the differential equation relating the output, V_1, to the input, V_2. No lengthy equation derivation was involved as in the path-vertex-elemental equation method. Furthermore, using the operational block diagram notation you learned in Chapter 2, equation (3.25) can be reduced to a single transfer function block as shown in Figure 3.6.

Now let's derive the equation for the circuit shown in Figure 3.4b using the impedance method. From the general impedance diagram shown in Figure 3.5b and equation (3.10), we can write the relationship between the output voltage and input current as

$$V_1 = \left(\frac{Z_1 Z_2}{Z_1 + Z_2}\right) i \qquad (3.28)$$

where from above

$$Z_1 = R \qquad \begin{matrix}(3.22)\\ \text{repeated}\end{matrix}$$

and

$$Z_2 = \frac{1}{CD} \qquad \begin{matrix}(3.23)\\ \text{repeated}\end{matrix}$$

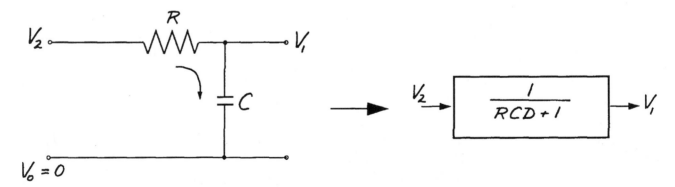

Figure 3.6. Transfer function block diagram equivalent for a resistor and a capacitor in series.

Substituting (3.22) and (3.23) into (3.28) gives

$$V_1 = \left(\frac{R\frac{1}{CD}}{R + \frac{1}{CD}} \right) i \qquad (3.29)$$

which can be simplified to

$$V_1 = \left(\frac{R}{RCD + 1} \right) i \qquad (3.30)$$

or

$$RCDV_1 + V_1 = Ri \qquad (3.31)$$

Here again we note that the operator D, or $d(\ )/dt$ is operating on the output voltage variable. So

$$RC\frac{dV_1}{dt} + V_1 = Ri \qquad (3.32)$$

Comparing (3.32) to (3.20) reveals that they are the same. So again we have been able to write the differential equation relating the output to the input without the need for the lengthy equation derivation we found using the path-vertex-elemental equation method.

Many electrical circuits you will encounter are simply combinations of elements arranged as voltage dividers or as elements in parallel. For example, the circuit shown in Figure 3.7a might look complicated, but it can easily be solved for the relationship between V_{out} and V_{in} in just a few steps as follows:

Step 1. Replace circuit elements with impedances as in Figure 3.7b.

Step 2. Combine parallel elements into a single equivalent element as in Figure 3.7c.

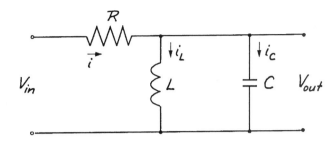

Figure 3.7(a). Electrical circuit containing all three elements.

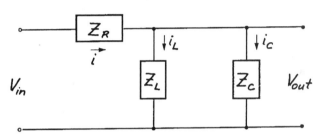

Figure 3.7(b). Replace elements with impedances.

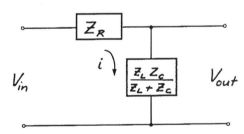

Figure 3.7(c). Reduce parallel elements to equivalent impedance and recognize reduced circuit as a voltage divider.

Step 3. Recognize the simplified circuit as a voltage divider and write the output-input equation by inspection in terms of impedances:

$$\frac{V_{out}}{V_{in}} = \frac{Z_{eq}}{Z_R + Z_{eq}} = \frac{\dfrac{Z_L Z_C}{Z_L + Z_C}}{Z_R + \dfrac{Z_L Z_c}{Z_L + Z_c}}$$

$$= \frac{Z_L Z_c}{Z_R Z_L + Z_R Z_c + Z_L Z_c}$$

Step 4. Substitute elemental impedances and perform algebra:

$$\frac{V_{out}}{V_{in}} = \frac{LD \times \dfrac{1}{CD}}{RLD + R \times \dfrac{1}{CD} + LD \times \dfrac{1}{CD}}$$

$$\frac{V_{out}}{V_{in}} = \frac{\dfrac{LD}{CD}}{\dfrac{RLCD^2 + R + LD}{CD}} = \frac{\dfrac{L}{R}D}{LCD^2 + \dfrac{L}{R}D + 1}$$

Step 5. Write differential equation by substituting $d(\ )/dt$ for the operator D:

$$(LC)\frac{d^2 V_{out}}{dt^2} + \left(\frac{L}{R}\right)\frac{dV_{out}}{dt} + V_{out} = \left(\frac{L}{R}\right)\frac{dV_{in}}{dt}$$

The final equation arrived at in Step 5 is a *second*-order ordinary linear differential equation with constant coefficients. It is called *second-order* because the second derivative of the output (dependent) variable is involved. You will study these types of equations later in Chapter 6. For now, look at what we have accomplished. In five easy steps, we derived the math model for a fairly complex electrical circuit. To prove to yourself just how easy and fast the impedance method is, stop now and try to derive the equation given at the end of step 5 using the path-vertex-elemental equation method.

Operational Block Diagram Method

Another tool that is very useful in visualizing how a system is working is the operational block diagram. In Chapter 2 we developed block diagrams for each of the three circuit elements and I stated then that it was best to develop these diagrams without using differentiator blocks. Now we will prepare a block diagram for the circuit of Figure 3.4.

When you develop a block diagram, first write the elemental equation for the circuit. We did this earlier, so from equation (3.12) we have

$$i = \frac{V_2 - V_1}{R} = \frac{1}{R}\left(V_2 - V_1\right)$$

or

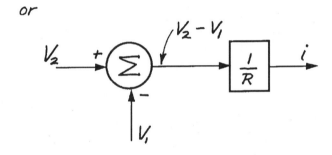

Figure 3.8.

Note how this block diagram was prepared. First you decide which variable is going to be the output. Then you solve for that variable and put it on the left side of the equation. Start drawing the diagram with the input or forcing variable, in this case V_2, at the far left of the diagram. Then just follow the equation drawing summers and multiplication blocks as needed.

The block diagram so far shows current as the output. We ultimately want V_1 as the output, so a block diagram for the capacitor is needed. Equation 3.13 is in the wrong form because it shows current as an output instead of an input. Following the steps, we rearrange the equation and draw the diagram.

$$\frac{dV_1}{dt} = \frac{1}{C}i$$

$$dV_1 = \frac{1}{C}i\,dt$$

$$V_1 = \int_0^t \left(\frac{1}{C}\right)i\,dt + (V_1)_{init}$$

or

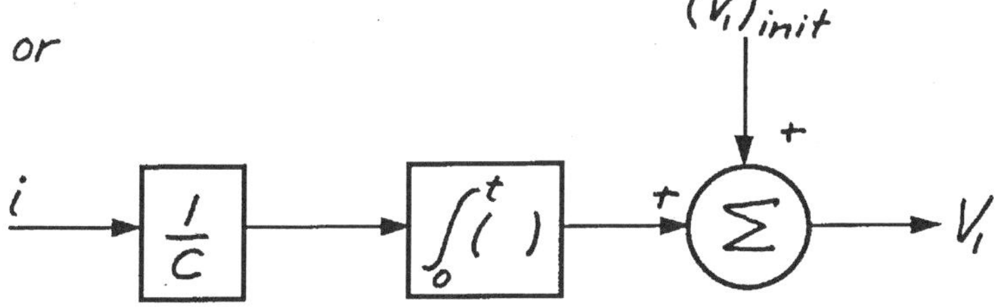

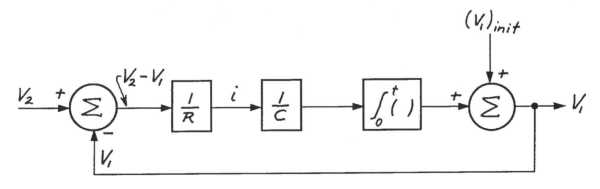

Figure 3.9.

Now we can add this to our previous diagram to give Figure 3.10. Notice how the output voltage V_1 on the right side "feeds back" and is subtracted from V_2. This creates a voltage difference, $V_2 - V_1$, which in turn causes a current to flow across the resistor. This current then flows

into the capacitor where it is integrated and produces the output voltage.

Look at how this circuit works as viewed by the block diagram. Start with no initial charge on the capacitor and the input voltage at zero. If a constant voltage V_2^* were suddenly applied to the circuit, the voltage difference out of the summer would equal V_2^* because V_1 is initially zero. A current equal to V_2^*/R will initially flow. The output voltage V_1 will not suddenly increase because the integrator takes time to convert its input into an output. After some time, Δt, a voltage $V_1^{(1)}$ will exist. This voltage is then subtracted from V_2^* to give a new current equal to $(V_2^* - V_1^{(1)})/R$. Again this new current, now lower than it was before, gets integrated during another time interval, Δt, and creates a higher output voltage $V_1^{(2)}$. Once again, the higher output voltage is fed back, where it reduces the current to the capacitor even more. The process continues until the current is reduced to zero. That occurs when $V_1 = V_2$. So you can see that a sudden change in the input voltage to this circuit passes through to the output, but only after the passage of time.

Figure 3.10.

If we let the initial charge on the capacitor $(V_1)_{init}$ be zero, use the operator $1/D$ for the integral, and replace RC with τ, then we can reduce the block diagram into that shown in Figure 3.11.

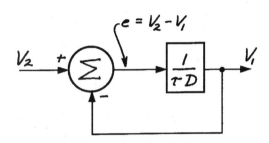

Figure 3.11.

From this block diagram we can write

$$e = V_2 - V_1$$

and

$$V_1 = \frac{1}{\tau D} e$$

(In feedback-control terminology, the symbol e is used to indicate error between the input and output.)

Combining these two equations to eliminate e gives

$$V_1 = \frac{1}{\tau D}(V_2 - V_1)$$

Rearranging gives

$$\tau D V_1 + V_1 = V_2$$

$$V_1(\tau D + 1) = V_2$$

$$V_1 = \frac{V_2}{\tau D + 1}$$

The block diagram for this last equation can be drawn as Figure 3.12.

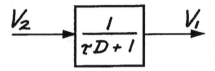

Figure 3.12.

As you can now see, the block diagram in Figure 3.10 can be reduced to the form shown in Figure 3.12. This latter diagram is often called the *transfer function* relating the output V_1 to the input V_2. The transfer function is expressed as output divided by input, or

$$\frac{V_1}{V_2} = \frac{1}{\tau D + 1}$$

Note that this equation is identical to equation 3.15b when τ is substituted for RC and D for $d(\ )/dt$ in (3.15b). You can see that if you can draw a block diagram, you can reduce it to the desired math model you're looking for.

Incidentally, the transfer function shown in Figure 3.12 is very common. You will encounter it many times in engineering systems. It is often called a *first-order lag* because it describes a first-order differential equation and because it causes the output to lag the input. We'll discuss the importance of this transfer function in more detail in the next chapter.

Free-body Diagram Method

You learned in Chapter 2 that symbolic (circuit) diagrams can be made for mechanical components. The "circuit" diagrams can be analyzed using the techniques presented in the previous two sections to arrive at math models for mechanical systems. You will now learn another powerful method for deriving equations for mechanical systems. It has its origins with Newton's laws of motion and involves the construction of *free-body diagrams*.

The free-body diagram method involves drawing a sketch of a mechanical component and labeling all of the forces and moments acting on the component. If the body is known to be at rest or is not accelerating, the sum of all the forces and the sum of all the moments acting on the body must equal zero. If the forces and moments acting on the body do not sum to zero, then the body will undergo translational and/or angular acceleration. Equations describing the motion of the body can be obtained by setting the sum of the forces equal to the mass of the body times the acceleration and the sum of the moments equal to the rotational inertia of the body times the angular acceleration.

Figure 3.13a shows a block with a mass m being pulled by a rope along a horizontal surface. Constructing a free-body diagram of the block involves isolating it in free space and showing all forces acting on the block. Next, choose a convenient axis so forces and moments can be summed and set either to zero, if no motion along that reference axis occurs, or to the mass (rotational inertia for mo-

ments) times the acceleration (linear for translation, angular for rotation). You must be careful to show *all* forces acting on the body, but *only* the forces acting on the body.

Figure 3.13b shows a correct free-body diagram of the block. The tension in the rope is identified as the force F. Note that the hand has nothing to do with the free-body diagram and that I assumed the tension in the rope acts horizontally to the surface. The force due to earth's gravity is shown as W and is equal to mg, where g is the gravitational constant. Since the block is not moving downward, there is an unknown force N acting at an unknown distance l which supports the block in the vertical direction. Since we know the block is sliding along the surface, we assume there is a resistive force f acting along the sliding surface of the block.

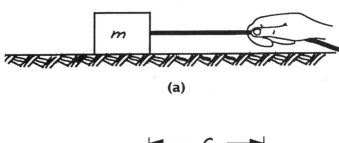

(a)

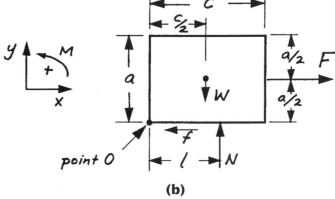

(b)

Figure 3.13. Constructing a free-body diagram.

We select a convenient set of axes and denote the positive direction as shown in Figure 3.13b. Then we write the equations in the three axes shown as follows:

Sum of forces in y direction equals zero (no motion in this axis)

$$\sum F_y = N - W = 0 \qquad (3.33)$$

Sum of forces in x direction equals mass times acceleration in x direction

$$\sum F_x = F - f = ma_x \qquad (3.34)$$

Sum of moments about point 0 equals zero (no rotational motion of block)

$$\sum M_0 = Nl - W\frac{c}{2} - F\frac{a}{2} = 0 \qquad (3.35)$$

You can see from (3.33) that the unknown force N is equal to the weight of the block. This information can be substituted into (3.35) and that equation solved for the unknown distance l as follows:

$$Wl = W\frac{c}{2} + F\frac{a}{2}$$

$$l = \frac{c}{2} + \frac{F}{W}\frac{a}{2} \qquad (3.36)$$

Equation (3.36) shows that the point of application of N increases as the force F increases. If l increases to the point where it becomes greater than c, the block will tip over.

Now let's examine (3.34) in more detail. The frictional force f is still unknown and another equation is needed to determine it. One possible assumption is that f is related to N. This assumption is often made when dry surfaces slide against one another. In this case we use a coefficient of friction μ and write

$$f = \mu N \qquad (3.37)$$

This allows us to determine f since we already know N is equal to W.

Another possible assumption is that the surfaces are lubricated causing the force f to be proportional to the velocity of the block v. We can then write

$$f = bv \qquad (3.38)$$

where b is the constant of proportionality. You will immediately recognize (3.38) as the equation for a damper (dashpot) discussed in Chapter 2.

Substituting (3.38) into (3.34) and expressing a_x as dv/dt gives

$$F - bv = m\frac{dv}{dt}$$

$$\frac{m}{b}\frac{dv}{dt} + v = \frac{1}{b}F \qquad (3.39)$$

Comparing (3.39) with (3.20) shows them to be identical to each other when the preferred electrical/mechanical analog discussed in Chapter 2 is used. The damper and the mass are therefore in parallel. The through variable F divides into two parts. One part overcomes the damper force and the other part accelerates the mass—you can see this clearly in (3.34).

3.3 First-Order Math Models

Now that you have tools for developing math models, let's use them to develop models for systems that are comprised of one or more energy dissipative elements and one energy storage element. We'll look at electrical systems first and then move on to translational and rotational mechanical systems, fluid systems, and thermal systems. As you read through this section, take the time to *practice* developing the math models as I suggest. I have found that the more circuits I analyze, the better I get at analyzing circuits and the longer the analysis techniques stay with me.

Electrical Math Models

Figure 3.14 shows five electrical circuits. Prepare the math models that describe the output variable as a function of the input variable as indicated by each circuit. Use any method you wish, but I strongly suggest that you use: (1) the path-vertex-elemental equation method; (2) the impedance method; or (3) the block diagram method. (Or, try all three if you feel ambitious!) Try not to look at my solutions until you have at least attempted to derive each of the math models.

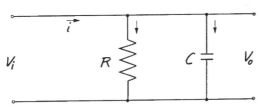

(a). *Prepare math model describing V_o as a function of i.*

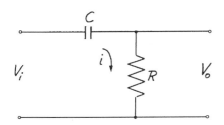

(b). *Prepare math model describing V_o as a function of V_i.*

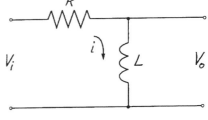

(c). *Prepare math model describing V_o as a function of V_i.*

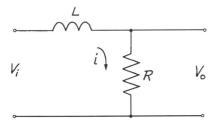

(d). *Prepare math model describing V_o as a function of V_i.*

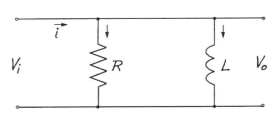

(e). *Prepare math model describing V_o as a function of V_i.*

Figure 3.14. Develop math models for these electrical systems.

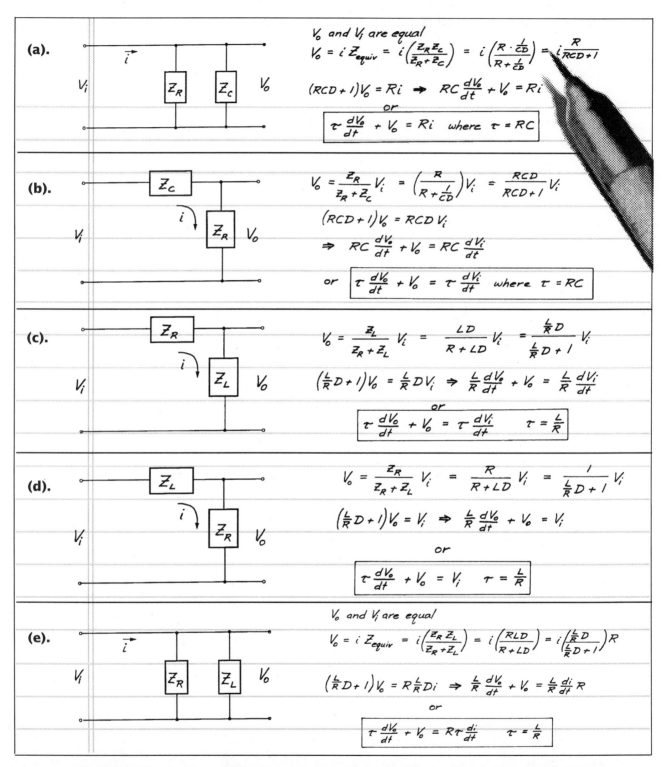

Figure 3.15. Math models for the electrical systems shown in Figure 3.14.

The method I prefer to use when analyzing electrical circuits is the impedance method. While the circuits shown in Figure 3.14 are relatively easy to analyze by either of the other two methods, those methods become very tedious when the circuits are more complex. The math models I developed and the ways in which they were developed are shown in Figure 3.15.

You should note several important common features that each of these models possess. First, they are all first-order ordinary linear differential equations with constant coefficients. Second, the left sides of all of the equations are identical except for the way the time constant is expressed. Finally, only the right, or input, side of each equation is different.

Mechanical Math Models

Figure 3.16 shows four mechanical systems to practice with. Prepare the math models that describe the output variable as a function of the input variable as indicated alongside each schematic. Use any method you wish, but I strongly suggest you use all four of the methods we discussed in this chapter. Once again, try not to look at my solutions until you have at least attempted to derive each of the math models.

The method I prefer to use when analyzing mechanical systems is the free-body diagram method. However, if the systems are very complicated, I frequently find myself reverting back to the impedance method. The math models I developed and the ways in which I developed them are shown in Figure 3.17.

Once again, note the common features that each of these models possess. They are all first-

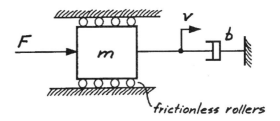

(a). Prepare math model describing the velocity v of the mass as a function of the force F.

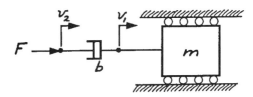

(b). Prepare math model describing the velocity v_1 as a function of the velocity v_2.

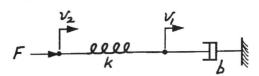

(c). Prepare math model describing the velocity v_1 as a function of the velocity v_2.

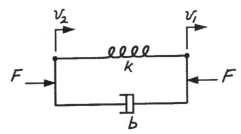

(d). Prepare math model describing the velocity differential v_{21} as a function of the force F.

Figure 3.16. Develop math models for these mechanical systems.

(a).

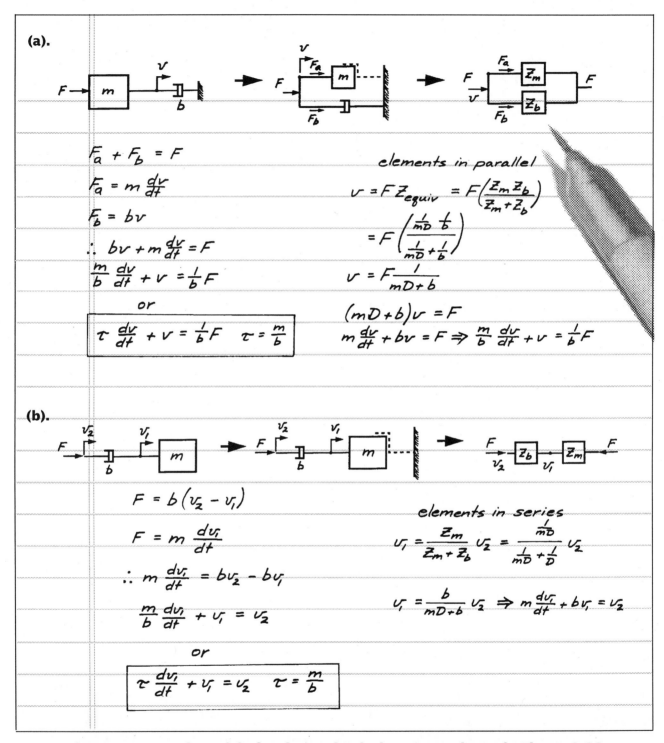

$$F_a + F_b = F$$

$$F_a = m \frac{dv}{dt}$$

$$F_b = bv$$

$$\therefore bv + m\frac{dv}{dt} = F$$

$$\frac{m}{b}\frac{dv}{dt} + v = \frac{1}{b}F$$

or

$$\boxed{\tau\frac{dv}{dt} + v = \frac{1}{b}F \qquad \tau = \frac{m}{b}}$$

elements in parallel

$$v = FZ_{equiv} = F\left(\frac{Z_m Z_b}{Z_m + Z_b}\right)$$

$$= F\left(\frac{\frac{1}{mD}\,b}{\frac{1}{mD} + \frac{1}{b}}\right)$$

$$v = F\frac{1}{mD + b}$$

$$(mD + b)v = F$$

$$m\frac{dv}{dt} + bv = F \Rightarrow \frac{m}{b}\frac{dv}{dt} + v = \frac{1}{b}F$$

(b).

$$F = b\left(v_2 - v_1\right)$$

$$F = m\frac{dv_1}{dt}$$

$$\therefore m\frac{dv_1}{dt} = bv_2 - bv_1$$

$$\frac{m}{b}\frac{dv_1}{dt} + v_1 = v_2$$

or

$$\boxed{\tau\frac{dv_1}{dt} + v_1 = v_2 \qquad \tau = \frac{m}{b}}$$

elements in series

$$v_1 = \frac{Z_m}{Z_m + Z_b}\,v_2 = \frac{\frac{1}{mD}}{\frac{1}{mD} + \frac{1}{D}}\,v_2$$

$$v_1 = \frac{b}{mD + b}\,v_2 \Rightarrow m\frac{dv_1}{dt} + bv_1 = v_2$$

Figure 3.17. Math models for the mechanical systems shown in Figure 3.16.

(continued on next page)

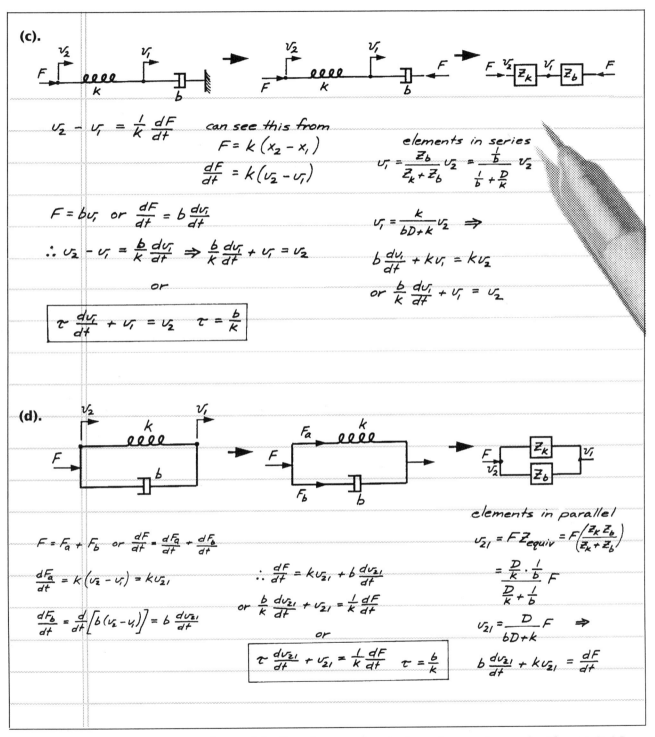

(c).

$$v_2 - v_1 = \frac{1}{k}\frac{dF}{dt} \quad \text{can see this from}$$
$$F = k(x_2 - x_1)$$
$$\frac{dF}{dt} = k(v_2 - v_1)$$

elements in series
$$v_1 = \frac{z_b}{z_k + z_b}v_2 = \frac{\frac{1}{b}}{\frac{1}{b}+\frac{D}{k}}v_2$$

$$F = bv_1 \quad \text{or} \quad \frac{dF}{dt} = b\frac{dv_1}{dt}$$

$$v_1 = \frac{k}{bD+k}v_2 \Rightarrow$$

$$\therefore v_2 - v_1 = \frac{b}{k}\frac{dv_1}{dt} \Rightarrow \frac{b}{k}\frac{dv_1}{dt} + v_1 = v_2$$

$$b\frac{dv_1}{dt} + kv_1 = kv_2$$

or

or $\frac{b}{k}\frac{dv_1}{dt} + v_1 = v_2$

$$\boxed{\tau \frac{dv_1}{dt} + v_1 = v_2 \quad \tau = \frac{b}{k}}$$

(d).

$$F = F_a + F_b \quad \text{or} \quad \frac{dF}{dt} = \frac{dF_a}{dt} + \frac{dF_b}{dt}$$

elements in parallel
$$v_{21} = FZ_{equiv} = F\left(\frac{z_k z_b}{z_k + z_b}\right)$$

$$\frac{dF_a}{dt} = k(v_2 - v_1) = kv_{21}$$

$$\therefore \frac{dF}{dt} = kv_{21} + b\frac{dv_{21}}{dt}$$

$$= \frac{\frac{D}{k}\cdot\frac{1}{b}}{\frac{D}{k}+\frac{1}{b}}F$$

$$\frac{dF_b}{dt} = \frac{d}{dt}\left[b(v_2 - v_1)\right] = b\frac{dv_{21}}{dt}$$

$$\text{or} \quad \frac{b}{k}\frac{dv_{21}}{dt} + v_{21} = \frac{1}{k}\frac{dF}{dt}$$

$$v_{21} = \frac{D}{bD+k}F \Rightarrow$$

or

$$\boxed{\tau\frac{dv_{21}}{dt} + v_{21} = \frac{1}{k}\frac{dF}{dt} \quad \tau = \frac{b}{k}}$$

$$b\frac{dv_{21}}{dt} + kv_{21} = \frac{dF}{dt}$$

Figure 3.17 *(continued).* **Math models for the mechanical systems shown in Figure 3.16.**

order ordinary linear differential equations with constant coefficients. The left sides are all identical except for the way the time constant is expressed. Only the right, or the input, side of each equation is different.

Now compare these equations with those that were developed for the electrical circuits. If you use the force-current and velocity-voltage analogy, you will see that the equations are essentially identical.

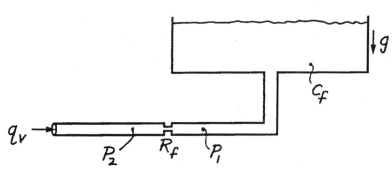

(a). Prepare math model describing pressure p_1 as a function of input pressure p_2.

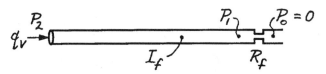

(b). Prepare math model describing: (1) q_v as a function of p_2; (2) p_1 as a function of p_2.

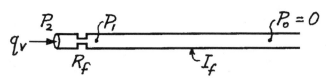

(c). Prepare math model describing: (1) q_v as a function of p_2; (2) p_1 as a function of p_2.

Figure 3.18. Develop math models for these fluid systems.

Fluid Math Models

Figure 3.18 shows three fluid systems for you to practice with. Prepare the math models that describe the output variable as a function of the input variable as indicated next to each schematic. Use any method you wish, but again I suggest that you use as many of the methods described as possible. Once again, try not to look at my solutions until you have at least attempted to derive each of the math models.

I really have no preferred method for analyzing fluid systems. If I get stuck, I revert back to a combination of the vertex-path-elemental equation method and the impedance method. The math models I developed and the ways in which I developed them are shown in Figure 3.19. Once again, note the common features that each of these models possess. They are all first-order ordinary linear differential equations with constant coefficients. The left sides are all identical except for the way the time constant is expressed. Only the right, or input, side of each equation is different.

Now compare these equations with those that were developed for the electrical circuits and the mechanical systems. If you use the force-current-flow rate and velocity-voltage-pressure analogy, you will see that the equations are essentially identical.

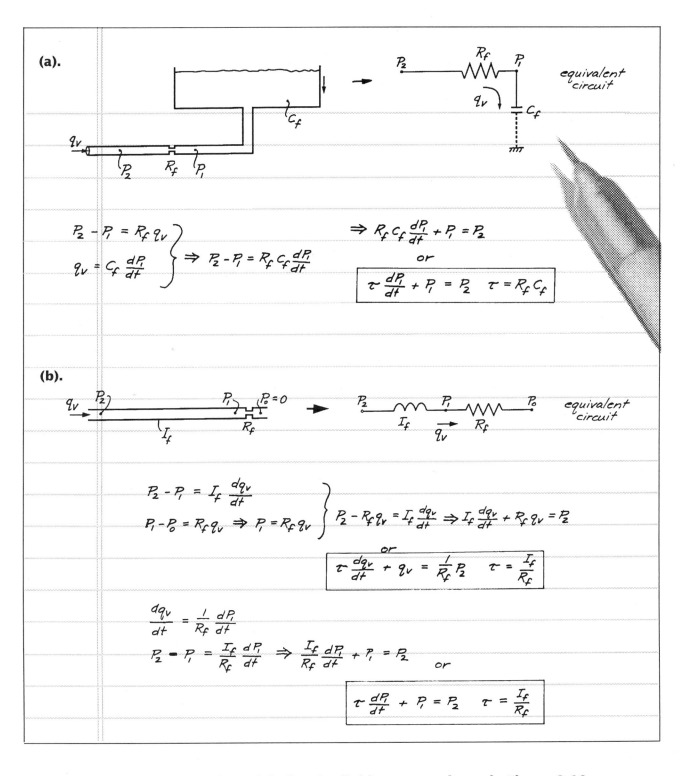

(a).

$$P_2 - P_1 = R_f q_v$$
$$q_v = C_f \frac{dP_1}{dt}$$
$$\Rightarrow P_2 - P_1 = R_f C_f \frac{dP_1}{dt}$$

$$\Rightarrow R_f C_f \frac{dP_1}{dt} + P_1 = P_2$$

or

$$\tau \frac{dP_1}{dt} + P_1 = P_2 \qquad \tau = R_f C_f$$

(b).

$$P_2 - P_1 = I_f \frac{dq_v}{dt}$$
$$P_1 - P_0 = R_f q_v \Rightarrow P_1 = R_f q_v$$
$$P_2 - R_f q_v = I_f \frac{dq_v}{dt} \Rightarrow I_f \frac{dq_v}{dt} + R_f q_v = P_2$$

or

$$\tau \frac{dq_v}{dt} + q_v = \frac{1}{R_f} P_2 \qquad \tau = \frac{I_f}{R_f}$$

$$\frac{dq_v}{dt} = \frac{1}{R_f} \frac{dP_1}{dt}$$

$$P_2 - P_1 = \frac{I_f}{R_f} \frac{dP_1}{dt} \Rightarrow \frac{I_f}{R_f} \frac{dP_1}{dt} + P_1 = P_2 \qquad or$$

$$\tau \frac{dP_1}{dt} + P_1 = P_2 \qquad \tau = \frac{I_f}{R_f}$$

Figure 3.19. Math models for the fluid systems shown in Figure 3.18.

(continued on next page)

(c).

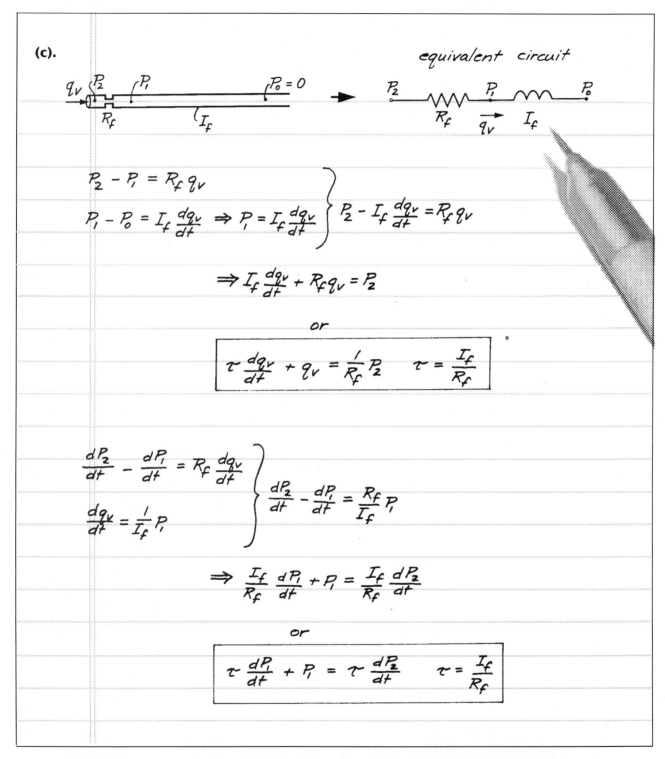

$$P_2 - P_1 = R_f q_v$$

$$P_1 - P_0 = I_f \frac{dq_v}{dt} \Rightarrow P_1 = I_f \frac{dq_v}{dt} \Bigg\} \quad P_2 - I_f \frac{dq_v}{dt} = R_f q_v$$

$$\Rightarrow I_f \frac{dq_v}{dt} + R_f q_v = P_2$$

or

$$\boxed{\tau \frac{dq_v}{dt} + q_v = \frac{1}{R_f} P_2 \qquad \tau = \frac{I_f}{R_f}}$$

$$\frac{dP_2}{dt} - \frac{dP_1}{dt} = R_f \frac{dq_v}{dt} \Bigg\} \quad \frac{dP_2}{dt} - \frac{dP_1}{dt} = \frac{R_f}{I_f} P_1$$

$$\frac{dq_v}{dt} = \frac{1}{I_f} P_1$$

$$\Rightarrow \frac{I_f}{R_f} \frac{dP_1}{dt} + P_1 = \frac{I_f}{R_f} \frac{dP_2}{dt}$$

or

$$\boxed{\tau \frac{dP_1}{dt} + P_1 = \tau \frac{dP_2}{dt} \qquad \tau = \frac{I_f}{R_f}}$$

Figure 3.19 *(continued).* **Math models for the fluid systems shown in Figure 3.18.**

Thermal Math Models

Figure 3.20 shows one thermal system for you to practice with. Prepare the math model that describes the output variable as a function of the input variable as indicated. Use any method you wish, but again I suggest that you use as many of the methods described as possible. Don't look at my solution until you have at least attempted to derive the math model.

I have no preferred method for analyzing thermal systems. Should I get stuck, I revert back to a combination of the vertex-path-elemental equation method and the impedance method. The math model I developed and the way in which I developed it is shown in Figure 3.21. Note that this is a first-order ordinary linear differential equations with constant coefficients.

Compare this equation with those that were developed for the electrical circuits, the me-

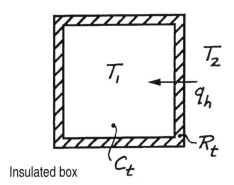

Prepare math model describing the temperature T_1 inside the box as a function of temperature T_2 outside the box.

Figure 3.20. Develop math model for this thermal system.

chanical systems, and the fluid system. If you use the force-current-flow rate-heat flow rate and velocity-voltage-pressure-temperature analogy, you will see that the equations are essentially identical.

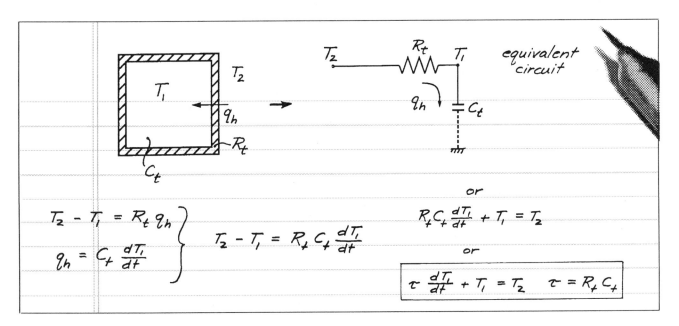

Figure 3.21. Math model for the thermal system shown in Figure 3.20.

Chapter

4 Analyzing First-Order Math Models

Objectives

When you have completed this chapter, you will be able to:

- **Solve first-order ordinary differential equations using both numerical and exact solution methods.**

- **Perform frequency analysis of first-order engineering systems and plot the frequency responses.**

- **Find solutions to first-order differential equations for step, ramp, pulse and arbitrary inputs.**

- **Perform a power analysis for systems responding to a sinusoidal input.**

4.1 Introduction

In the previous chapter you discovered that systems modeled as a combination of an ideal energy storage element and an ideal energy dissipative element lead to first-order linear ordinary differential equations with constant coefficients. You found out that this differential equation usually must be "solved" in order to obtain the time-varying response of the system to a time-varying input forcing function.

You will discover in this chapter that the solution provides a great deal of information about the behavior of the real system being modeled. You will first learn a simple approximation method of solution that can be used to solve any ordinary differential equation. This method is extremely powerful and does not require a great deal of mathematical skill. It will be used to introduce you to the various types of input forcing functions commonly encountered when modeling engineering systems.

Once you understand the basic nature of the solutions to differential equations, you will then learn how to obtain the exact solutions. You will discover that exact solutions provide more insight into the system being modeled than do approximate solutions. However, exact methods do involve more mathematical manipulations.

The first exact solution method you will learn involves separating variables so the differential equation can be directly integrated. Following this you will learn a more methodical approach that takes advantage of the fact that the differential equation is linear and has constant coefficients. You will discover that this method is easy and can be used to obtain exact

solutions to higher-order linear differential equations with constant coefficients.

You will also be introduced in this chapter to so-called "frequency response" or "frequency domain" solutions and analysis. You will find these extremely important methods of analysis are nothing more than the solution of the differential equation to a sinusoidal input function.

I can't emphasize enough just how important first-order linear ordinary differential equations with constant coefficients are to engineering and science. Probably 80% of all engineering systems you will ever encounter can be completely described by, or contain at least one component that can be described by, one of these equations. Make friends with them—they will serve you well!

4.2 Response to a Step Input

In this section, we will investigate the behavior of an engineering system that can be described by a first-order linear ordinary differential equation with constant coefficients in response to a sudden change in the input. The sudden change of the input variable from one level to another is called a *step* input. The step input is an approximation of many real-world inputs to engineering systems. For example, when we tramp on a car's accelerator or flip on a switch that applies voltage to a circuit, we're applying a step input. We want to find out how the system responds to such an input—that is, we want to find the output as a function of time. You will first learn how to numerically solve the equation subjected to a step input. Then you will learn how to find the exact solution.

Numerical Solution Method

Let's begin solving first-order math models using a model we developed in Chapter 3 that described the electrical circuit in Figure 3.4. The circuit and math model are repeated in Figure 4.1 and equation (4.1).

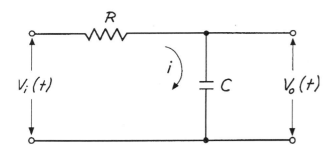

Figure 4.1.

$$RC\frac{dV_o}{dt}+V_o=V_i \qquad (4.1)$$

New labels for the voltages are used to clearly indicate which one is the input voltage, V_i, and which is the output voltage V_o. If we had the solution to this differential equation, $V_o(t)$, we could use it to determine: (1) how various values of R and C affect the output voltage for a given input voltage; and (2) how the output voltage varies with different input voltage functions, given values for R and C.

First I'll show you an easy method for solving differential equations. It is called the "numerical solution" method and it can easily be set up on a spreadsheet or by writing a simple BASIC, FORTRAN, or C computer program. Take the

math model as given by equation (4.1) and re-write it as

$$RC\frac{\Delta V_o}{\Delta t}+V_o=V_i \qquad (4.2)$$

In this equation, $\Delta V_o/\Delta t$ is an approximation for the derivative, dV_o/dt. Now solve for ΔV_o

$$\Delta V_o=\left[\frac{1}{RC}(V_i-V_o)\right]\Delta t \qquad (4.3)$$

Equation (4.3) tells us that if we know V_i at every point in time and if we know the initial value of V_o, then we can use this equation to determine the change in the output voltage ΔV_o that occurs in the time interval Δt. We can therefore solve equation (4.1) using a time stepping process. That is, we start at time $t = 0$ where we know V_o and V_i. Then we take a time step Δt and use equation (4.3) to compute the *change* in the output voltage that occurs during this time interval. We add this change to the previous value of V_o to get the new V_o. The process is continued until we reach some point in time where the variables are no longer changing or where we are no longer interested in the results.

The algorithm for numerically solving a differential equation is very straightforward. One is provided for a first-order differential equation in Table 4.1. Notice in this table that selecting a value for the time step is arbitrary. I've selected a value equal to 1/10th of the constant $\tau = RC$. We'll investigate this selection later but for now keep in mind that the size of the time step affects the speed of computing the solution and its accuracy.

Table 4.1.
Algorithm for numerically solving a first-order differential equation.

Given:

(1) V_o at $t = 0$ (that is, the initial condition of the dependent variable)

(2) V_i as a function of time (that is, the forcing function at any point in time)

(3) Values for R and C

Step 1 Initialize variables:
$$\tau = RC$$
$$t = 0$$
$$t_{end} = 5\tau$$
$$\Delta t = \tau/10$$
$$V_o = (V_o)_{init}$$

Step 2 Increment time and check if done
$$t = t + \Delta t$$
If $t = t_{end}$ then stop.

Step 3 Compute $V_i(t)$ from the given function

Step 4 Solve for ΔV_o
$$\Delta V_o = \left[\frac{1}{RC}(V_i - V_o)\right]\Delta t$$

Step 5 Determine new V_o
$$V_o = V_o + \Delta V_o$$

Step 6 Go back to **Step 2**

Spreadsheets are very useful in solving differential equations numerically, as math models can be set up very quickly. An Excel® spreadsheet implementing the algorithm given in Table 4.1 is shown in Table 4.2. Lines 1 through 7 accept the given data ($R = 1$, $C = 1$, $(V_o)_{init} = 0$ and $V_i(t) = 10$) and compute the time step. Note that I have chosen R and C so their product is equal to unity. That is, the time constant of the circuit is equal to 1 second. Line 9 provides a label for the results showing t, V_i, V_o and ΔV_o. Lines 10 through the end implement the iterative solution technique.

The results from this spreadsheet are shown in Table 4.3, and Figure 4.2 shows an Excel graph of the results. You can see that a constant voltage, suddenly applied at $t = 0$, does not produce an instantaneous output. The output voltage builds rapidly at first, having an initial rate of increase of 1 volt per 0.1 seconds (10 volts per second). Since V_o is initially zero, you can see that the initial rate is from equation (4.3):

$$\frac{\Delta V_o}{\Delta t} = \left[\frac{1}{RC}(V_i - V_o)\right] = \left[\frac{1}{1}(10-0)\right] = 10$$
$$(4.4)$$

The rate slows as the output voltage builds. You will recall from the block diagram discussion of this circuit given in Section 3.3 that this is due to the output voltage feeding back and reducing the current flowing into the capacitor. As this current is reduced, the charge on the capacitor asymptotically builds to the value of the input voltage. Note also that in 1 second the output voltage is 6.51 volts, or 65.1% of its final value.

Table 4.2. First-order step response spreadsheet implementation.

	A	B	C	D
1	R	1		
2	C	1		
3	Tau	=B1*B2		
4	del t	=B3/10		
5	(Vo)init	0		
6	Vi*	10		
7	Tend	=5*B3		
8				
9	t	Vi	Vo	del Vo
10	0	=$B$6	=B5	=(1/$B$3*(B10-C10))*$B$4
11	=A10+$B$4	=$B$6	=C10+D10	=(1/$B$3*(B11-C11))*$B$4
12	=A11+$B$4	=$B$6	=C11+D11	=(1/$B$3*(B12-C12))*$B$4
13	=A12+$B$4	=$B$6	=C12+D12	=(1/$B$3*(B13-C13))*$B$4
14	=A13+$B$4	=$B$6	=C13+D13	=(1/$B$3*(B14-C14))*$B$4
15	=A14+$B$4	=$B$6	=C14+D14	=(1/$B$3*(B15-C15))*$B$4
16	=A15+$B$4	=$B$6	=C15+D15	=(1/$B$3*(B16-C16))*$B$4
17	=A16+$B$4	=$B$6	=C16+D16	=(1/$B$3*(B17-C17))*$B$4
18	=A17+$B$4	=$B$6	=C17+D17	=(1/$B$3*(B18-C18))*$B$4
19	=A18+$B$4	=$B$6	=C18+D18	=(1/$B$3*(B19-C19))*$B$4
20	=A19+$B$4	=$B$6	=C19+D19	=(1/$B$3*(B20-C20))*$B$4
21	=A20+$B$4	=$B$6	=C20+D20	=(1/$B$3*(B21-C21))*$B$4
22	=A21+$B$4	=$B$6	=C21+D21	=(1/$B$3*(B22-C22))*$B$4
23	=A22+$B$4	=$B$6	=C22+D22	=(1/$B$3*(B23-C23))*$B$4
24	=A23+$B$4	=$B$6	=C23+D23	=(1/$B$3*(B24-C24))*$B$4
25	=A24+$B$4	=$B$6	=C24+D24	=(1/$B$3*(B25-C25))*$B$4
26	=A25+$B$4	=$B$6	=C25+D25	=(1/$B$3*(B26-C26))*$B$4
27	=A26+$B$4	=$B$6	=C26+D26	=(1/$B$3*(B27-C27))*$B$4
28	=A27+$B$4	=$B$6	=C27+D27	=(1/$B$3*(B28-C28))*$B$4
29	=A28+$B$4	=$B$6	=C28+D28	=(1/$B$3*(B29-C29))*$B$4
30	=A29+$B$4	=$B$6	=C29+D29	=(1/$B$3*(B30-C30))*$B$4
31	=A30+$B$4	=$B$6	=C30+D30	=(1/$B$3*(B31-C31))*$B$4
32	=A31+$B$4	=$B$6	=C31+D31	=(1/$B$3*(B32-C32))*$B$4
33	=A32+$B$4	=$B$6	=C32+D32	=(1/$B$3*(B33-C33))*$B$4
34	=A33+$B$4	=$B$6	=C33+D33	=(1/$B$3*(B34-C34))*$B$4
35	=A34+$B$4	=$B$6	=C34+D34	=(1/$B$3*(B35-C35))*$B$4
36	=A35+$B$4	=$B$6	=C35+D35	=(1/$B$3*(B36-C36))*$B$4
37	=A36+$B$4	=$B$6	=C36+D36	=(1/$B$3*(B37-C37))*$B$4
38	=A37+$B$4	=$B$6	=C37+D37	=(1/$B$3*(B38-C38))*$B$4
39	=A38+$B$4	=$B$6	=C38+D38	=(1/$B$3*(B39-C39))*$B$4
40	=A39+$B$4	=$B$6	=C39+D39	=(1/$B$3*(B40-C40))*$B$4
41	=A40+$B$4	=$B$6	=C40+D40	=(1/$B$3*(B41-C41))*$B$4
42	=A41+$B$4	=$B$6	=C41+D41	=(1/$B$3*(B42-C42))*$B$4
43	=A42+$B$4	=$B$6	=C42+D42	=(1/$B$3*(B43-C43))*$B$4
44	=A43+$B$4	=$B$6	=C43+D43	=(1/$B$3*(B44-C44))*$B$4
45	=A44+$B$4	=$B$6	=C44+D44	=(1/$B$3*(B45-C45))*$B$4
46	=A45+$B$4	=$B$6	=C45+D45	=(1/$B$3*(B46-C46))*$B$4

Table 4.3. First-order step response results.

	A	B	C	D
1	R	1		
2	C	1		
3	Tau	1		
4	del t	0.1		
5	(Vo)init	0		
6	Vi*	10		
7	Tend	5		
8				
9	t	Vi	Vo	del Vo
10	0	10	0.00	1.00
11	0.10	10	1.00	0.90
12	0.20	10	1.90	0.81
13	0.30	10	2.71	0.73
14	0.40	10	3.44	0.66
15	0.50	10	4.10	0.59
16	0.60	10	4.69	0.53
17	0.70	10	5.22	0.48
18	0.80	10	5.70	0.43
19	0.90	10	6.13	0.39
20	1.00	10	6.51	0.35
21	1.10	10	6.86	0.31
22	1.20	10	7.18	0.28
23	1.30	10	7.46	0.25
24	1.40	10	7.71	0.23
25	1.50	10	7.94	0.21
26	1.60	10	8.15	0.19
27	1.70	10	8.33	0.17
28	1.80	10	8.50	0.15
29	1.90	10	8.65	0.14
30	2.00	10	8.78	0.12
31	2.10	10	8.91	0.11
32	2.20	10	9.02	0.10
33	2.30	10	9.11	0.09
34	2.40	10	9.20	0.08
35	2.50	10	9.28	0.07
36	2.60	10	9.35	0.06
37	2.70	10	9.42	0.06
38	2.80	10	9.48	0.05
39	2.90	10	9.53	0.05
40	3.00	10	9.58	0.04
41	3.10	10	9.62	0.04
42	3.20	10	9.66	0.03
43	3.30	10	9.69	0.03
44	3.40	10	9.72	0.03
45	3.50	10	9.75	0.03
46	3.60	10	9.77	0.02

In 2 seconds, it's 8.78 volts or 87.8%, and in 3 seconds it reaches 9.58 volts, or 95.8% of the final value.

The response of the circuit shown in Figure 4.2 is called a step response. The input forcing function takes a step at time $t = 0$ from its value of 0 at $t < 0$ to a value of 10 volts. There are other types of input forcing functions and we will discuss these later. For now, let's keep the input function equal to a step and change its value to 5 volts. The results are shown in Figure 4.3. Note that the output voltage appears to have exactly the same shape as it did in Figure

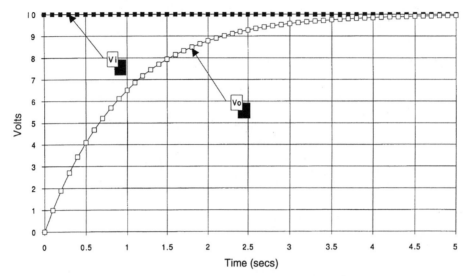

Figure 4.2. Numerical solution of equation (4.1) math model to a step change in input voltage

(Vi = 10, R = 1, C = 1, Tau = RC = 1, delt = 0.1).

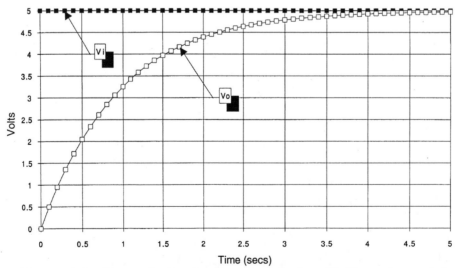

Figure 4.3. Numerical solution of equation (4.1) math model to a step change in input voltage

(Vi = 5, R = 1, C = 1, Tau = RC = 1, delt = 0.1) .

4.2. Also note that at the end of 1 second the output voltage is approximately 3.25 volts, or 65.1% of the input voltage. This is the same percentage we found in Figure 4.2.

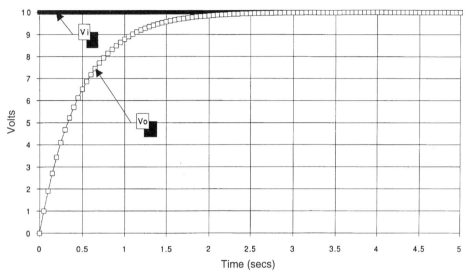

Figure 4.4. Numerical solution of equation (4.1) math model to a step change in input voltage

(Vi = 10, R = 0.5, C = 1, Tau = RC = 0.5, delt = 0.05).

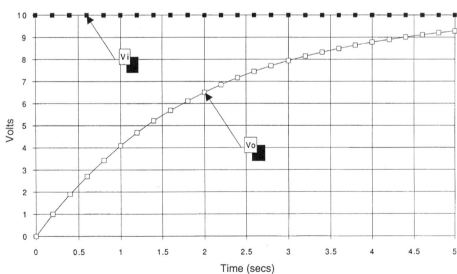

Figure 4.5. Numerical solution of equation (4.1) math model to a step change in input voltage

(Vi = 10, R = 2, C = 1, Tau = RC = 2, delt = 0.2).

Now let's put the value of the step input voltage back to 10 volts and vary the values of R and C. Before we do this, look at equation (4.4). When the product of R and C is small, the initial rate of change of the output voltage will be large. Also, it is the *product* of R and C that affects this rate. Thus, if R were to decrease to 0.5 and C increase to 2, the product would remain equal to unity. Figures 4.4 and 4.5 show the voltage output when τ = 0.5 and 2.0, respectively. You can see that when τ is small, the output voltage reaches the input voltage level more rapidly. Note in both of these figures that when the output voltage reaches approximately 65% of the input, the time is equal to τ. That's because the time constant is a *characteristic* of first-order linear ordinary differential equations. It controls the form of the relationship between the input and the output. When we investigate exact solutions of first-order differential equations you will find the exact values given in Table 4.4 hold for the step responses.

Note that the results we obtained for these percentages are not quite equal to the exact values listed in Table 4.4. That's because the numerical solution is approximate and not exact. The accuracy of the solution is determined to a large extent by the size of the time step used in the solution. Table 4.5 and Figure 4.6 show the results previously shown in Table 4.3 and Figure 4.2, but with the time step changed to 0.05 seconds ($\tau/20$). You can see from the table of results that the output voltage is now equal to 6.42 volts, or 64.2% of the input voltage when t = 1.0 seconds. As Δt is made smaller, the solution approaches the exact value.

Table 4.4.

**Step response times
for first-order
linear differential equations.**

Time	Output as a percentage of input
1 τ	63.2
2 τ	86.5
3 τ	95.0
4 τ	98.2
5 τ	99.3
6 τ	99.8

Table 4.5.
**First-order step response results
(delt = 0.05 seconds).**

	A	B	C	D
1	R	1		
2	C	1		
3	Tau	1		
4	del t	0.05		
5	(Vo)init	0		
6	Vi*	10		
7	Tend	5		
8				
9	t	Vi	Vo	del Vo
10	0	10	0.00	0.50
11	0.05	10	0.50	0.48
12	0.10	10	0.98	0.45
13	0.15	10	1.43	0.43
14	0.20	10	1.85	0.41
15	0.25	10	2.26	0.39
16	0.30	10	2.65	0.37
17	0.35	10	3.02	0.35
18	0.40	10	3.37	0.33
19	0.45	10	3.70	0.32
20	0.50	10	4.01	0.30
21	0.55	10	4.31	0.28
22	0.60	10	4.60	0.27
23	0.65	10	4.87	0.26
24	0.70	10	5.12	0.24
25	0.75	10	5.37	0.23
26	0.80	10	5.60	0.22
27	0.85	10	5.82	0.21
28	0.90	10	6.03	0.20
29	0.95	10	6.23	0.19
30	1.00	10	6.42	0.18
31	1.05	10	6.59	0.17
32	1.10	10	6.76	0.16
33	1.15	10	6.93	0.15
34	1.20	10	7.08	0.15
35	1.25	10	7.23	0.14
36	1.30	10	7.36	0.13
37	1.35	10	7.50	0.13
38	1.40	10	7.62	0.12
39	1.45	10	7.74	0.11
40	1.50	10	7.85	0.11
41	1.55	10	7.96	0.10
42	1.60	10	8.06	0.10
43	1.65	10	8.16	0.09
44	1.70	10	8.25	0.09
45	1.75	10	8.34	0.08
46	1.80	10	8.42	0.08

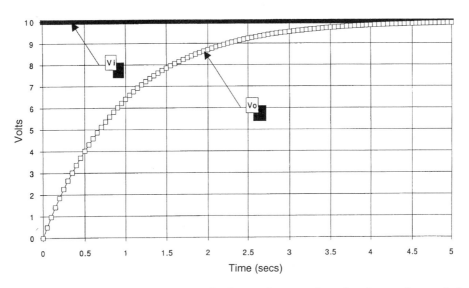

Figure 4.6. Numerical solution of equation (4.1) math model to a step change in input voltage (Vi = 10, R = 1, C = 1, Tau = RC = 1, delt = 0.05).

Exact Solution Method

You've seen how easy it is to solve differential equations approximately. Now I want to show you how to obtain the exact solution.

First write the math model given in equation (4.1) in the form

$$\tau \frac{dV_o}{dt} + V_o = V_{is} \qquad (4.5)$$

where V_{is} equals the value of the step input function as shown in Figure 4.7.

Since V_{is} is a constant, we can divide both sides of equation (4.5) by V_{is} and obtain the fol-

lowing *nondimensional* version of the equation

$$\tau \frac{d\left(\dfrac{V_o}{V_{is}}\right)}{dt} + \left(\frac{V_o}{V_{is}}\right) = 1 \qquad (4.6)$$

I hope this doesn't confuse you. All I have done is divide the output voltage by the *constant* input voltage. Since both have units of volts, (V_o / V_{is}) has units of volts/volt, which is dimensionless.

Note that the new dependent variable is now (V_o / V_{is}) instead of just V_o. This will make the solution of the equation easier. Also note that the step input is now a *unit step* input,

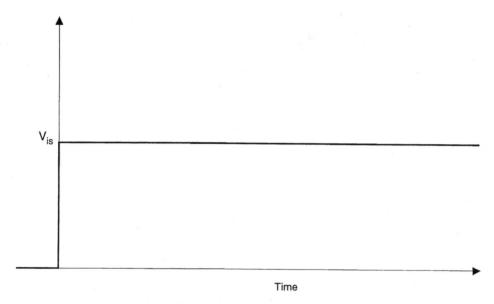

Figure 4.7. The step input function.

which is commonly used in linear system analysis. Let $(V_o / V_{is}) = V$ and rewrite (4.6) as

$$\tau \frac{dV}{dt} + V = 1 \qquad (4.7)$$

This is identical to (4.5) except the output variable is dimensionless and the input is now a unit step function.

It is possible to solve equation (4.7) by direct integration. The method of solution is called *separation of variables*. You arrange the equation so all variables involving V are on one side of the equation and all involving t are on the other. First write (4.7) as

$$\tau \frac{dV}{dt} = 1 - V \qquad (4.8)$$

Now divide both sides by $1-V$ and multiply both sides by dt/τ

$$\frac{dV}{1-V} = \frac{1}{\tau} dt \qquad (4.9)$$

The variables have been successfully separated since only terms containing V are on the left side and only terms containing t are on the right side. You can now integrate both sides of the equation

$$\int_0^V \frac{dV}{1-V} = \frac{1}{\tau} \int_0^t dt \qquad (4.10)$$

The integral on the right side of the equation is easy to solve, but you may have to look up the one on the left side in a table of integrals. The result is

$$-\ln(1-V) = \frac{1}{\tau}t$$

or

$$\ln(1-V) = -\frac{t}{\tau} \qquad (4.11)$$

Now take the antilog of both sides of (4.11)

$$e^{\ln(1-V)} = e^{-t/\tau}$$
$$1-V = e^{-t/\tau}$$
$$V = 1-e^{-t/\tau} \qquad (4.12)$$

Equation (4.12) is the exact solution of equation (4.7). We can get it back to dimensional values if necessary simply by substituting (V_o / V_{is}) for V. That is

$$V = \frac{V_o}{V_{is}} = 1-e^{-t/\tau}$$

or

$$V_o = V_{is}\left(1-e^{-t/\tau}\right) \qquad (4.13)$$

This is the exact solution of equation (4.5).

One of the great advantages of an exact solution is that it provides you with more insight into the behavior of the system than the numerical solution. For example, both equations (4.12) and (4.13) make it very clear that the output voltage is a function only of (t / τ). Since

τ has units of time, (t / τ) has units of seconds/second—that is, dimensionless time. Equation (4.12) is a completely general solution that is valid for all values of τ. Equation (4.13) makes it clear that the output voltage is directly proportional to the input voltage. *This is a distinguishing characteristic of all linear differential equations and is why they are called linear.* That is, the output is directly proportional to the input.

Another great advantage an exact solution has over a numerical solution is you can precisely calculate the response of the system for any value of a step input voltage at any point in time. Figure 4.8 shows a plot of equation (4.12) and Figure 4.9 shows a plot of equation (4.13) with $\tau = 1$ and $V_{is} = 10$ volts. Figure 4.8 shows it all very concisely. *Regardless* of the values selected for τ and V_{is}, this graph shows the solution. (This is another advantage of non-dimensionalizing, as we did in equation (4.6).) Figure 4.9, on the other hand, shows a particular solution for $\tau = 1$ second and $V_{is} = 10$ volts.

Let's go back now and determine how accurate our numerical solution was as a function of the size of the time step. You will recall that we solved the equation with time steps equal to $\tau / 10$ and $\tau / 20$. Table 4.6 shows the exact solution versus the numerical solution using time steps equal to $\tau / 5$, $\tau / 10$, and $\tau / 20$. You can see that acceptable accuracy is achieved using a time step equal to 1/10th of the time constant.

You might be wondering at this time why we bother with numerical solutions if exact solutions to our math models can be obtained. The answer is, if you can obtain an exact solution, do so. It is always easier and more insightful to work with exact solutions. Unfortunately, it's not always as easy as you saw here to find exact solutions to differential equations. Sometimes it takes more time to find a solution than it does to solve the equations numerically. Furthermore, numerically solving equations also works for arbitrary input functions and nonlinear differential equations. It is very difficult, and sometimes impossible, to find exact solutions to nonlinear differential equations. Of course you will have to experiment with the size of the time step when solving nonlinear differential equations to make sure your solution is sufficiently accurate for your purposes. That is the major drawback with using numerical methods. However, there are numerical methods that can automatically adjust the time step and there are better methods of numerically solving differential equations than the one I presented above. Two good references on this subject are provided at the end of this chapter.

Take some time now and study Figure 4.10. Learn to recog-

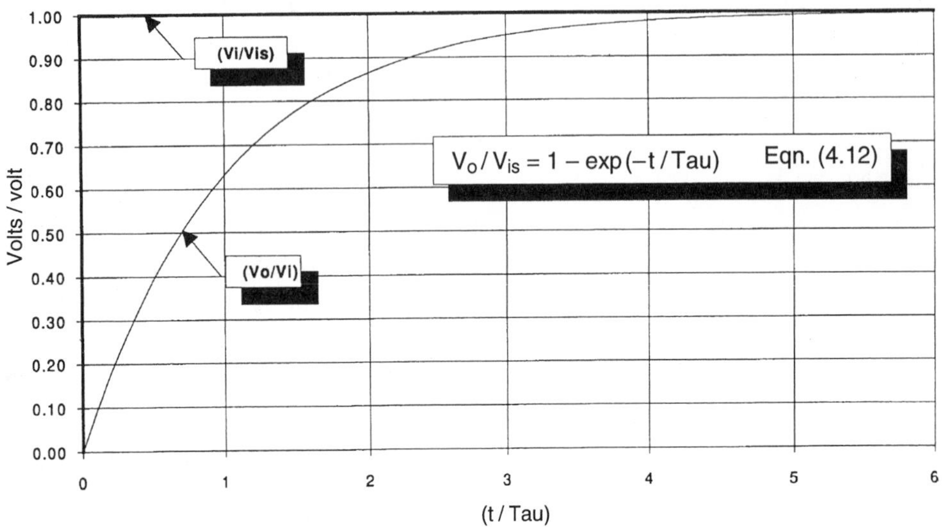

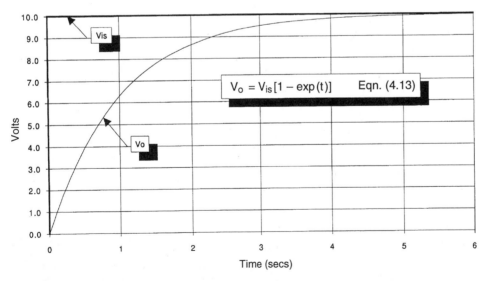

$$V_O / V_{is} = 1 - \exp(-t / Tau) \qquad \text{Eqn. (4.12)}$$

Figure 4.8. Exact dimensionless solution of equation (4.1) math model to a dimensionless step change in input voltage.

$$V_O = V_{is} [1 - \exp(t)] \qquad \text{Eqn. (4.13)}$$

Figure 4.9. Exact solution of equation (4.1) math model to a step change in input voltage (Vi = 10 volts, Tau = 1 sec).

Table 4.6.
Accuracy of numerical solution of a differential equation.

Time	Output Voltage (volts)			
	$\Delta t = \tau/5$	$\Delta t = \tau/10$	$\Delta t = \tau/20$	Exact
0.0τ	0.00	0.00	0.00	0.00
0.2τ	2.00	1.90	1.85	1.81
0.4τ	3.60	3.44	3.37	3.30
$0.6t$	4.88	4.69	4.60	4.51
0.8τ	5.90	5.70	5.60	5.51
1.0τ	6.72	6.51	6.42	6.32
1.2τ	7.38	7.18	7.08	6.99
1.4τ	7.90	7.71	7.62	7.53
1.6τ	8.32	8.15	8.06	7.98
1.8τ	8.66	8.50	8.42	8.35
2.0τ	8.93	8.78	8.71	8.65
2.2τ	9.14	9.02	8.95	8.89
2.4τ	9.31	9.20	9.15	9.09
2.6τ	9.45	9.35	9.31	9.26
2.8τ	9.56	9.48	9.43	9.39
3.0τ	9.65	9.58	9.54	9.50

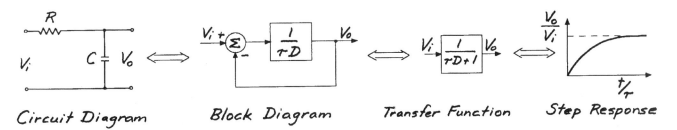

Circuit Diagram   Block Diagram   Transfer Function   Step Response

Figure 4.10.

nize first-order differential equations and visualize their symbolic representations and solutions to a step response. Remember that this important class of equations represents about 80% of the dynamics you will encounter in engineering system design and analysis.

4.3 Response to a Sinusoidal Input

You will now learn about so-called *frequency analysis* and *frequency response* of engineering systems. This basically means investigating the response of the system to sinusoidal inputs of different frequencies. This is an extremely important method of analysis, as poor frequency response can be a drawback in many types of engineering systems. For example, a telephone that attenuates high frequencies too much does not deliver intelligible speech. In an instrumentation system, if the sensor and amplifier don't have a good frequency response, false measurements may result. Frequency analysis is actually very easy to understand, but for some reason is often misunderstood.

We just investigated the response of our system to a step input. That is, in equation (4.1) we let $V_i(t)$ equal a constant V_{is} at $t = 0$. The system equation and its step response were found approximately and exactly.

Now you will investigate the response of the same system to a sinusoidal input. That is, the input to equation (4.1) will be of the form

$$V_i = V_{is} \sin(2\pi f t) \qquad (4.14)$$

This is a sinusoid with a maximum amplitude of V_{is} and a frequency of f cycles per second (hertz or Hz). Often you will see equation (4.14) expressed in terms of *circular frequency*, denoted by the symbol ω. That is,

$$V_i = V_{is} \sin(\omega t) \qquad (4.15)$$

where $\omega = 2\pi f$.

You will also sometimes see sinusoids expressed in terms of their *period*, denoted by the symbol T. That is,

$$V_i = V_{is} \sin\left(2\pi \frac{t}{T}\right) \qquad (4.16)$$

where $T = 1/f = 2\pi/\omega$.

Any of these forms are fine to use and Figure 4.11 shows all of these relationships.

Numerical Solution Method

We will first obtain the response of the system to the sinusoidal input by numerically integrating the differential equation. Solve the following equation

$$\tau \frac{\Delta V_o}{\Delta t} + V_o = V_{is} \sin 2\pi f t \qquad (4.17)$$

where $V_{is} = 10$ volts, $(V_o)_{init} = 0$, $\tau = 1$ second, and $f = 1$ Hz,

and use $\Delta t = \tau/20$.

I obtained the results shown in Figure 4.12 using the Excel spreadsheet given in Table 4.7. There are several important things to note in this figure. The response V_o appears to have a frequency identical to the input, but it is shifted in time so it lags behind the input voltage V_i. The maximum amplitude of the output is also less than the maximum amplitude of the input. Notice also that there is an initial start-up transient, during which the output voltage tries to catch up with the input voltage but never quite makes it. After a while, this transient appears to die away.

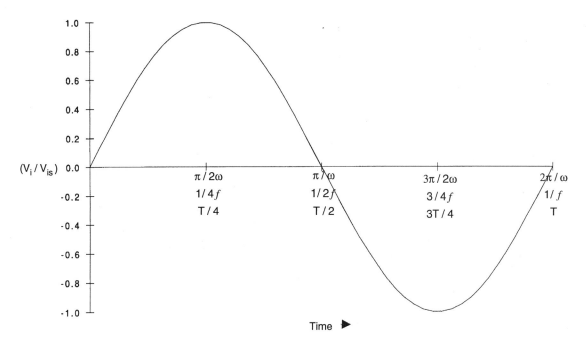

Figure 4.11. Various ways to express a sinusoid.

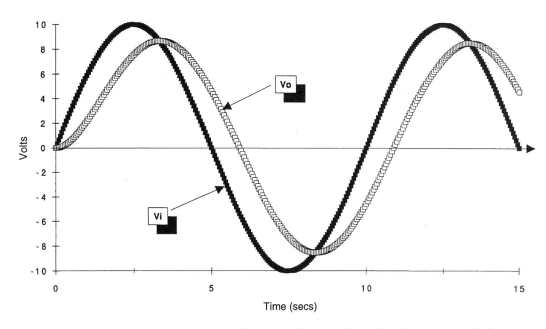

Figure 4.12. Numerical solution of equation (4.1) math model to a sinusoidal input voltage (f = 0.1, Tau = 1, delt = 0.05).

Table 4.7.
First-order sin response spreadsheet implementation.

	A	B	C	D
1	R	1		
2	C	1		
3	Tau	=B1*B2		
4	del t	=B3/20		
5	(Vo)init	0		
6	Vi*	10		
7	f	0.5		
8	Tend	=5*B3		
9				
10	t	Vi	Vo	del Vo
11	0	=$B$6*SIN(2*PI()*$B$7*A11)	=B5	=(1/$B$3*(B11-C11))*$B$4
12	=A11+$B$4	=$B$6*SIN(2*PI()*$B$7*A12)	=C11+D11	=(1/$B$3*(B12-C12))*$B$4
13	=A12+$B$4	=$B$6*SIN(2*PI()*$B$7*A13)	=C12+D12	=(1/$B$3*(B13-C13))*$B$4
14	=A13+$B$4	=$B$6*SIN(2*PI()*$B$7*A14)	=C13+D13	=(1/$B$3*(B14-C14))*$B$4
15	=A14+$B$4	=$B$6*SIN(2*PI()*$B$7*A15)	=C14+D14	=(1/$B$3*(B15-C15))*$B$4
16	=A15+$B$4	=$B$6*SIN(2*PI()*$B$7*A16)	=C15+D15	=(1/$B$3*(B16-C16))*$B$4
17	=A16+$B$4	=$B$6*SIN(2*PI()*$B$7*A17)	=C16+D16	=(1/$B$3*(B17-C17))*$B$4
18	=A17+$B$4	=$B$6*SIN(2*PI()*$B$7*A18)	=C17+D17	=(1/$B$3*(B18-C18))*$B$4
19	=A18+$B$4	=$B$6*SIN(2*PI()*$B$7*A19)	=C18+D18	=(1/$B$3*(B19-C19))*$B$4
20	=A19+$B$4	=$B$6*SIN(2*PI()*$B$7*A20)	=C19+D19	=(1/$B$3*(B20-C20))*$B$4
21	=A20+$B$4	=$B$6*SIN(2*PI()*$B$7*A21)	=C20+D20	=(1/$B$3*(B21-C21))*$B$4
22	=A21+$B$4	=$B$6*SIN(2*PI()*$B$7*A22)	=C21+D21	=(1/$B$3*(B22-C22))*$B$4
23	=A22+$B$4	=$B$6*SIN(2*PI()*$B$7*A23)	=C22+D22	=(1/$B$3*(B23-C23))*$B$4
24	=A23+$B$4	=$B$6*SIN(2*PI()*$B$7*A24)	=C23+D23	=(1/$B$3*(B24-C24))*$B$4
25	=A24+$B$4	=$B$6*SIN(2*PI()*$B$7*A25)	=C24+D24	=(1/$B$3*(B25-C25))*$B$4
26	=A25+$B$4	=$B$6*SIN(2*PI()*$B$7*A26)	=C25+D25	=(1/$B$3*(B26-C26))*$B$4
27	=A26+$B$4	=$B$6*SIN(2*PI()*$B$7*A27)	=C26+D26	=(1/$B$3*(B27-C27))*$B$4
28	=A27+$B$4	=$B$6*SIN(2*PI()*$B$7*A28)	=C27+D27	=(1/$B$3*(B28-C28))*$B$4
29	=A28+$B$4	=$B$6*SIN(2*PI()*$B$7*A29)	=C28+D28	=(1/$B$3*(B29-C29))*$B$4
30	=A29+$B$4	=$B$6*SIN(2*PI()*$B$7*A30)	=C29+D29	=(1/$B$3*(B30-C30))*$B$4
31	=A30+$B$4	=$B$6*SIN(2*PI()*$B$7*A31)	=C30+D30	=(1/$B$3*(B31-C31))*$B$4
32	=A31+$B$4	=$B$6*SIN(2*PI()*$B$7*A32)	=C31+D31	=(1/$B$3*(B32-C32))*$B$4
33	=A32+$B$4	=$B$6*SIN(2*PI()*$B$7*A33)	=C32+D32	=(1/$B$3*(B33-C33))*$B$4
34	=A33+$B$4	=$B$6*SIN(2*PI()*$B$7*A34)	=C33+D33	=(1/$B$3*(B34-C34))*$B$4
35	=A34+$B$4	=$B$6*SIN(2*PI()*$B$7*A35)	=C34+D34	=(1/$B$3*(B35-C35))*$B$4
36	=A35+$B$4	=$B$6*SIN(2*PI()*$B$7*A36)	=C35+D35	=(1/$B$3*(B36-C36))*$B$4
37	=A36+$B$4	=$B$6*SIN(2*PI()*$B$7*A37)	=C36+D36	=(1/$B$3*(B37-C37))*$B$4
38	=A37+$B$4	=$B$6*SIN(2*PI()*$B$7*A38)	=C37+D37	=(1/$B$3*(B38-C38))*$B$4
39	=A38+$B$4	=$B$6*SIN(2*PI()*$B$7*A39)	=C38+D38	=(1/$B$3*(B39-C39))*$B$4
40	=A39+$B$4	=$B$6*SIN(2*PI()*$B$7*A40)	=C39+D39	=(1/$B$3*(B40-C40))*$B$4
41	=A40+$B$4	=$B$6*SIN(2*PI()*$B$7*A41)	=C40+D40	=(1/$B$3*(B41-C41))*$B$4
42	=A41+$B$4	=$B$6*SIN(2*PI()*$B$7*A42)	=C41+D41	=(1/$B$3*(B42-C42))*$B$4
43	=A42+$B$4	=$B$6*SIN(2*PI()*$B$7*A43)	=C42+D42	=(1/$B$3*(B43-C43))*$B$4
44	=A43+$B$4	=$B$6*SIN(2*PI()*$B$7*A44)	=C43+D43	=(1/$B$3*(B44-C44))*$B$4
45	=A44+$B$4	=$B$6*SIN(2*PI()*$B$7*A45)	=C44+D44	=(1/$B$3*(B45-C45))*$B$4
46	=A45+$B$4	=$B$6*SIN(2*PI()*$B$7*A46)	=C45+D45	=(1/$B$3*(B46-C46))*$B$4

I repeated the numerical solution using the same input amplitude, initial condition, and time constant, while increasing the frequency of the input to 0.5, 1, and 2 Hz. The results are shown in Figures 4.13 through 4.15. You can see that the amplitude of the output signal decreased each time the frequency was increased and the input signal lagged further behind the input signal.

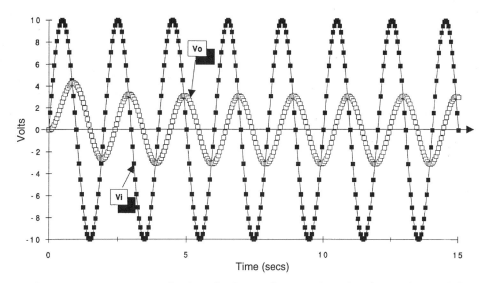

Figure 4.13. Numerical solution of equation (4.1) math model to a sinusoidal input voltage (f = 0.5, Tau = 1, delt = 0.05).

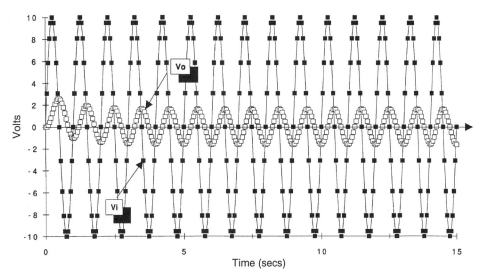

Figure 4.14. Numerical solution of equation (4.1) math model to a sinusoidal input voltage (f = 1, Tau = 1, delt = 0.05).

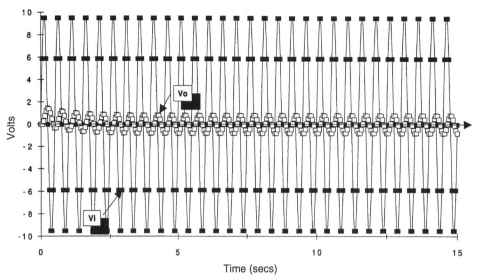

Figure 4.15. Numerical solution of equation (4.1) math model to a sinusoidal input voltage (f = 2, Tau = 1, delt = 0.05).

You can compute a so-called *signal attenuation factor* by dividing the maximum amplitude of the output signal by the maximum amplitude of the input signal. It is given the symbol A_r. If these attenuation factors are then plotted against the corresponding value of the input frequency, a graph similar to that shown in Figure 4.16 results. You can see that this system passes low-frequency input signals without too much attenuation, but it definitely attenuates the high-frequency signals. This system is often called a *first-order low-pass filter*. Many engineering systems other than electrical circuits have exactly the same type of frequency response characteristic shown in this figure. You can see this characteristic in a ship responding to waves, for example. Very high-frequency waves cause very little ship motion, but long-period (low-frequency) waves can cause a lot of motion.

You can also compute the amount the output signal leads or lags behind the input signal and plot this against the corresponding frequency. The amount the output signal leads or lags the input signal is generally shown in degrees and is called the *phase angle*. One complete cycle of a sinusoid is 360 degrees, so a lead is shown as 0 to +180 degrees and a lag as 0 to −180 degrees. In our system the output signal is lagging behind the input signal. This is called *phase lag*, and we can compute the angle by measuring the number of seconds the output signal lags behind the input, dividing this by the number of seconds for half of a cycle and multiplying by 180 degrees. Figure 4.17 shows the approximate results for this system.

The plots shown in Figure 4.16 and 4.17 are called the *frequency response of the system*. In several engineering fields Figure 4.16 is called the *response amplitude operator* (or *RAO* for short) and Figure 4.17 is called the *phase angle plot*. Regardless of their name, they show in a summary form: (1) the *ratio* of the output and input signals as a function of frequency or period of the input signal; and (2) the *phase shift* between the output and input signals as a function of frequency or period of the input signal. Many times you will see these plots shown using log scales. (I'll explain why in the next section.)

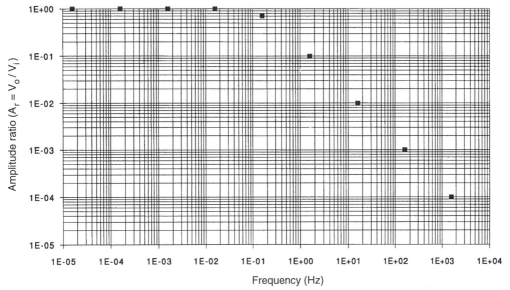

Figure 4.16. Estimated ratio of output and input voltages for equation (4.1) math model (Tau = 1).

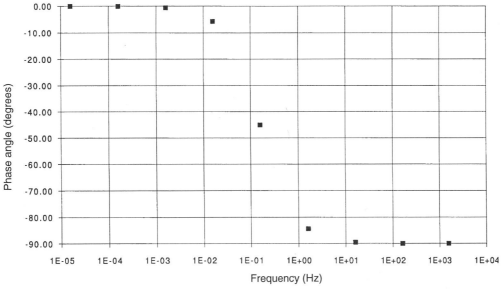

Figure 4.17. Estimated phase angle between output and input voltages for equation (4.1) math model (Tau = 1).

Exact Solution Method

I'll now explain how to determine the exact solution to a differential equation with a sinusoidal input. Before I do this, however, I'm going to digress in order to show you a more general way to solve differential equations. Study this method carefully, as we will be using it for the remainder of this book.

A characteristic of a linear ordinary differential equation of any order is its adherence to the *superposition principle*. That means if $G_1(t)$ and $G_2(t)$ are the outputs of a linear system in response to inputs $F_1(t)$ and $F_2(t)$, respectively, then an output corresponding to a *linear combination* of inputs such as $C_1F_1(t) + C_2F_2(t)$ [C_1 and C_2 are constants] is the linear combination $C_1G_1(t) + C_2G_2(t)$. This is really what is meant when the adjective *linear* is used to describe a first-order *linear* ordinary differential equation with constant coefficients. The superposition principle of linear systems is extremely important.

The solution of any linear differential equation is made up of two additive parts. One part is called the *homogeneous solution* and the other the *particular solution*. When the differential equation is arranged in its proper form with the response or output variable and its derivatives on the left side of the equal sign and the input variable and its derivatives on the right side, then the homogeneous solution is associated with the left side of the equation and the particular solution with the right. The left side of a differential equation is often called the *characteristic equation* and the right side the *forcing function*.

The *homogeneous solution* is found by setting the characteristic equation to zero. That is, we solve the differential equation as if there were no input. This may at first seem odd but, as you will soon discover, it's easy to find this solution and the solution is a fundamental characteristic of the differential equation and the system it represents.

For example, in our system presently under study, the characteristic equation is

$$\tau \frac{dV_o}{dt} + V_o$$

We set this to zero in order to obtain the homogeneous solution. That is

$$\tau \frac{dV_o}{dt} + V_o = 0 \qquad (4.18)$$

The homogeneous solution always has at least one solution of the form

$$V_o = Ae^{rt} \qquad (4.19)$$

where A and r are unknown values. We can differentiate this solution once to get dV_o/dt. Then we can substitute the assumed solution for V_o into (4.18) giving

$$\tau A r e^{rt} + Ae^{rt} = 0 \qquad (4.20)$$

Dividing both sides by Ae^{rt} results in

$$\tau r + 1 = 0 \qquad (4.21)$$

This equation must hold for all time if (4.19) is to be a solution to (4.18) for all time. That is, (4.21) requires that

$$r = -\frac{1}{\tau} \qquad (4.22)$$

and gives the homogeneous solution as

$$V_{oH} = Ae^{-t/\tau} \qquad (4.23)$$

The constant A is still unknown and must be determined from the initial conditions after the particular solution is found, but (4.23) is a solution of (4.18).

Several methods are available for finding the *particular solution* to linear ordinary differential equations with constant coefficients. We'll use the so-called *method of undetermined coefficients*. This method involves finding a function that is similar in appearance to the input or forcing function, but which contains undetermined coefficients. The undetermined coefficients are found by substituting the function into the differential equation and determining the coefficients which allow both sides of the differential equation to remain equal for all time.

For example, in the case of a step input function we can use an undetermined coefficient that is a constant B. That is,

$$V_{oP} = B \qquad (4.24)$$

Then we substitute this particular solution into our system math model and get

$$\tau\frac{dB}{dt} + B = V_{is} \qquad (4.25)$$

Since the derivative of a constant is zero, dB/dt is zero and we are left with

$$B = V_{is}$$

which holds for all time. Thus, the particular solution is

$$V_{oP} = V_{is} \qquad (4.26)$$

The complete solution is obtained by adding the homogeneous and particular solutions. That is,

$$V_o = V_{oH} + V_{oP}$$

or

$$V_o = Ae^{-t/\tau} + V_{is} \qquad (4.27)$$

We can now determine the value for the coefficient A using the initial condition

$$V_o = 0 \text{ at } t = 0 \qquad (4.28)$$

Substituting (4.28) into (4.27) gives

$$0 = Ae^{-0/\tau} + V_{is} = A + V_{is}$$

or

$$A = -V_{is}$$

The final solution then is equal to

$$V_o = V_{is} - V_{is}e^{-t/\tau}$$

or

$$V_o = V_{is}\left(1 - e^{-t/\tau}\right) \qquad (4.29)$$

which is the same as we obtained previously using the method of separation of variables.

The method of undetermined coefficients for finding a particular solution can be generalized as follows. The forcing function $F(t)$ will always have the general form

$$F(t) = e^{at}\cos bt\left(p_m t^m + p_{m-1}t^{m-1} + \cdots + p_0\right)$$
$$+ e^{at}\sin bt\left(q_m t^m + q_{m-1}t^{m-1} + \cdots + q_0\right)$$

where some of the constants a, b, p_0, p_m, ... q_0, q_m, ... q_0 may be zero. Then a particular solution to the differential equation of form similar to that of $F(t)$ is

$$y_p(t) = e^{at}\cos bt\left(k_m t^m + k_{m-1}t^{m-1} + \cdots + k_0\right)$$
$$+ e^{at}\sin bt\left(l_m t^m + l_{m-1}t^{m-1} + \cdots + l_0\right)$$

The coefficients k_m, k_{m-1}, ... k_0, l_m, l_{m-1}, ... l_0 are determined by substituting $y_p(t)$ into the differential equation and choosing the coefficients so the equation remains equal on both sides for all time.

For example, let's find the particular solution for the forcing function

$$V_i = V_{is}\sin 2\pi ft$$

We will use as the general solution

$$V_{oP} = k_o\cos 2\pi ft + l_o\sin 2\pi ft \qquad (4.30)$$

where k_o and l_o must be determined. Substituting (4.30) into our system math model

$$\tau\frac{dV_o}{dt} + V_o = V_{is}\sin 2\pi ft$$

gives

$$\tau\frac{d}{dt}\left(k_o\cos 2\pi ft + l_o\sin 2\pi ft\right)$$
$$+ \left(k_o\cos 2\pi f + l_o\sin 2\pi ft\right)$$
$$= V_{is}\sin 2\pi ft$$

Carrying out the differentiation and collecting like terms gives

$$\left(\tau 2\pi fk_o + l_o\right)\cos 2\pi ft + \left(k_o - \tau 2\pi fl_o\right)\sin 2\pi ft$$
$$= V_{is}\sin 2\pi ft \qquad (4.31)$$

We determine the coefficients k_o and l_o so (4.31) is satisfied for all time. For this to happen, the coefficient for the cosine term must be zero and the coefficient for the sine term must equal V_{is}. That means the following must be true for all time

$$\tau 2\pi fk_o + l_o = 0 \qquad (4.32)$$

and

$$k_o - \tau 2\pi fl_o = V_{is} \qquad (4.33)$$

Equations (4.32) and (4.33) are two equations in two unknowns, k_o and l_o. The simultaneous solution of these algebraic equations gives

$$k_o = \frac{V_{is}}{1 + \left(\tau 2\pi f\right)^2}$$

$$l_o = -\frac{\tau 2\pi f V_{is}}{1 + \left(\tau 2\pi f\right)^2}$$

Substituting these constants into (4.30) gives the particular solution as

$$V_{oP} = \frac{V_{is}}{1 + \left(\tau 2\pi f\right)^2}\sin 2\pi ft$$

$$-\frac{\tau 2\pi f V_{is}}{1 + \left(\tau 2\pi f\right)^2}\cos 2\pi ft \qquad (4.34)$$

Now add the homogeneous solution given by equation (4.23) and the particular solution given by equation (4.34) to get the complete solution

$$V_o = Ae^{-t/\tau} + \frac{V_{is}}{1 + \left(\tau 2\pi f\right)^2}\sin 2\pi ft$$

$$-\frac{\tau 2\pi f V_{is}}{1 + \left(\tau 2\pi f\right)^2}\cos 2\pi ft \qquad (4.35)$$

Obtain A using the initial condition (V_o = 0 at t = 0). That is,

$$V_o = 0 = A - \frac{\tau 2\pi f V_{is}}{1 + \left(\tau 2\pi f\right)^2}$$

or

$$A = \frac{\tau 2\pi f}{1 + \left(\tau 2\pi f\right)^2}V_{is}$$

The exact solution then is

$$\frac{V_o}{V_{is}} = \underbrace{\frac{\tau 2\pi f}{1 + \left(\tau 2\pi f\right)^2}e^{-t/\tau}}_{\text{transient}}$$

$$+\underbrace{\frac{1}{1 + \left(\tau 2\pi f\right)^2}\sin 2\pi ft - \frac{\tau 2\pi f}{1 + \left(\tau 2\pi f\right)^2}\cos 2\pi ft}_{\text{nontransient}}$$

$$(4.36)$$

Note the first term in this exact solution is the transient part noticed when we solved the equation numerically. After around $t = 3\tau$ seconds, this part of the solution approaches zero because of the exponential term. We are then left with the nontransient part of the solution which lasts for all time.

The *nontransient* part of the solution can also be placed in the following simpler form

$$\frac{V_o}{V_{is}} = C\sin\left(2\pi ft + \varphi\right) \qquad (4.37)$$

where   C is the amplitude attenuation factor and φ is the phase lag.

To do this, all we need are a few trigonometric identities and algebra.

Recall the trigonometric identity

$$C\sin\left(\varphi + \theta\right) = C\cos\varphi\sin\theta + C\sin\varphi\cos\theta$$

$$(4.38)$$

Compare (4.38) with the nontransient part of (4.36). You can see they are identical if we let

$$\theta = 2\pi f t \qquad (4.39)$$

$$C \cos\varphi = \frac{1}{1 + (\tau 2\pi f)^2} \qquad (4.40)$$

$$C \sin\varphi = -\frac{(\tau 2\pi f)}{1 + (\tau 2\pi f)^2} \qquad (4.41)$$

The angle φ is the phase lag angle. We can obtain an equation for it by dividing (4.41) by (4.40). That is,

$$\frac{C \sin\varphi}{C \cos\varphi} = \tan\varphi = -(\tau 2\pi f)$$

or

$$\varphi = \tan^{-1}(-\tau 2\pi f) \qquad (4.42)$$

You can obtain C using another trigonometric identity

$$\sin^2\varphi + \cos^2\varphi = 1 \qquad (4.43)$$

Substituting (4.40) and (4.41) into (4.43) and solving for C gives

$$C = \frac{1}{\sqrt{1 + (\tau 2\pi f)^2}} \qquad (4.44)$$

Now we rewrite (4.37) using (4.42) and (4.44)

$$V_o = \frac{V_{is}}{\sqrt{1 + (\tau 2\pi f)^2}} \sin\left[2\pi f t + \tan^{-1}(-\tau 2\pi f)\right]$$

$$(4.45)$$

Equation (4.45) is the exact solution to our system math model given in (4.1) for a sinusoidal input whose maximum amplitude is V_{is} and whose frequency is f, after the initial transient has died out. The equation shows that the maximum amplitude of the output voltage is dependent on the frequency, as well as the amplitude, of the input signal. The equation also shows that the phase lag between the input and output signals is dependent on the frequency of the input signal. Also note that the time constant, τ, plays a key role in determining the frequency response. The time constant appears in the equation as a multiplier of frequency. This makes it behave as a scaling factor for the frequency.

We can use (4.45) to calculate the ratio of the maximum output amplitude to the maximum input amplitude, as well as the phase shift between the two signals, as a function of frequency. Table 4.8 shows the results. The frequency is listed in terms of the time constant and is increased each entry by a multiple of 10 (an *order of magnitude* increase) so that a broad range of frequencies is covered. If $t = 1$ second as we have been using, the frequency range in the table goes from 0.0000159 Hz to 159,000 Hz. For some unknown reason it is still customary, particularly when presenting frequency response data of electrical components, to show the amplitude ratio using an acoustics unit called decibels. A decibel, abbreviated dB, is simply 20 times the base 10 log of the amplitude ratio; that is,

$$dB = 20 \log\left|\left(\frac{V_o}{V_{is}}\right)\right|$$

If you have a gut feel for decibels, then use them. Otherwise, just use the base 10 log of the ratio. The amplitude ratio is given in the table as a decimal ratio and in dB.

The data in Table 4.8 is plotted in Figure 4.16. Log-log paper is used for two reasons. First, it allows the large range of frequencies to be accommodated. Second, it shows that at a frequency around 1 Hz, the amplitude ratio starts decreasing at a nearly constant rate. The frequency at which this occurs is called the *break* or *corner* frequency. For a first-order circuit such as this, the break frequency is defined as

$$f_{break} = \frac{1}{2\pi\tau}$$

Note in Table 4.8 that, at this frequency, the amplitude ratio is 0.707 (–3 dB) and the phase angle is –45°. Quite often, experimentally determined frequency response data for electrical, mechanical, hydraulic, and thermal components are presented as graphs like Figure 4.16. The break point can be determined as that frequency where the amplitude ratio is .707 or –3 dB.

The low-pass characteristic of our system is clearly shown in Figure 4.16. It is customary to call the frequency range from zero to the break frequency the *pass band*, and the region above the break frequency the *stop band*. Such terms are taken from the field of analog filter design, but are equally applicable to any engineering system. But take care—a voltage input signal to this circuit at the break frequency ($f = 1/2\pi\tau$) is being attenuated nearly 30%. Furthermore, the phase shift between the input and output voltage signal is 45 degrees. Note also that for every decade increase in the frequency beyond the break frequency, the amplitude ratio decreases by a factor of 10 (or –20 dB per decade). Also note that the phase shift approaches 90° as the frequency approaches infinity.

The frequency response characteristic shown in Figure 4.16 completely describes our system. If you have the frequency response of a system and it looks like Figure 4.16, then you should be able to visualize the step response, the

Table 4.8.
First-order exact frequency response.

frequency	frequency	Ar (ratio)	20*log(Ar)	Phase Lag
	(Tau=1)	volts/volt	db	deg
0.0001/(2*pi*Tau)	1.59E-05	1.00E+00	-4.34E-08	-0.01
0.001/(2*pi*Tau)	1.59E-04	1.00E+00	-4.34E-06	-0.06
0.01/(2*pi*Tau)	1.59E-03	1.00E+00	-4.34E-04	-0.57
0.1/(2*pi*Tau)	1.59E-02	9.95E-01	-4.32E-02	-5.71
1/(2*pi*Tau)	1.59E-01	7.07E-01	-3.01E+00	-45.00
10/(2*pi*Tau)	1.59E+00	9.95E-02	-2.00E+01	-84.29
100/(2*pi*Tau)	1.59E+01	1.00E-02	-4.00E+01	-89.43
1000/(2*pi*Tau)	1.59E+02	1.00E-03	-6.00E+01	-89.94
10,000/(2*pi*Tau)	1.59E+03	1.00E-04	-8.00E+01	-89.99
100,000/(2*pi*Tau)	1.59E+04	1.00E-05	-1.00E+02	-90.00
1,000,000/(2*pi*Tau)	1.59E+05	1.00E-06	-1.20E+02	-90.00

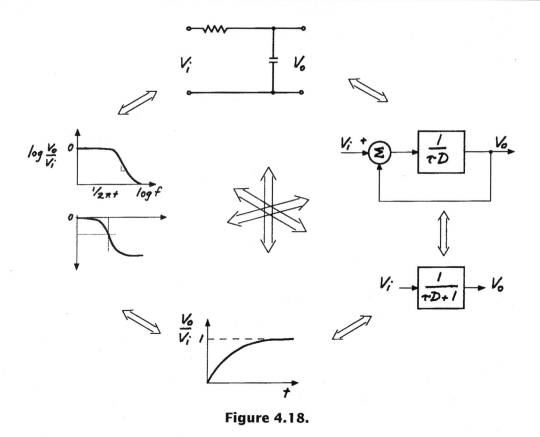

Figure 4.18.

transfer function, etc. Figure 4.18 shows all of the tools we used in analyzing our math model for this system. Study this figure and try to keep in mind the relationships for a system described by a first-order differential equation.

One last thing about the frequency response of systems. While I have presented this method in mathematical terms, the frequency response of any real system or component can often be obtained experimentally, as follows: The system is tested with the input being a sinusoid of variable amplitude and frequency. The output amplitude and phase shift is then measured. The ratio of the maximum output and maximum input amplitudes, and the phase shift between the two signals, are plotted against frequency. The results completely characterize the system.

Quite often math models can be built of systems using experimental frequency response data when it is difficult or impossible to prepare a math model of the components. For example, electrical engineers frequently excite an electrical component with an input sinusoid of constant amplitude and variable frequency and then measure the output amplitude as a function of the input frequency. The electrical component can be fully characterized in this manner. In the same manner, ocean engineers often characterize the frequency response of floating ocean structures experimentally by conducting model tests in an ocean wave laboratory in which the magnitude and frequency of the waves (input) are varied and the response of the vessel (output) is measured.

4.4 Response to Other Input Functions

You've now studied the response of a first-order linear ordinary differential equation with constant coefficients to a step input and to a sinusoidal input. The responses of first-order systems to these two types of input forcing functions are very important and you must gain a working knowledge of them.

Needless to say, there are a great variety of input functions that can be described mathematically and the first-order system response solutions to these functions can be obtained. We will look at a couple more in this section to give you additional insight into the behavior of first-order systems.

Response to a Ramp Input

A step input to a system can be severe. Harking back to our automotive example, it's much easier on your car when you press the accelerator down slowly and consistently than when you jam your foot to the floor. Many real systems are able to handle a step input without damage, but others are not. A gentler input may be needed to study the behavior of the system. A ramp input is such a function. To obtain a ramp, the input $V_i(t)$ to equation (4.1) is made

equal to

$$V_i(t) = \dot{V}_{ir} \times t \qquad (4.46)$$

where $\dot{V}_{ir}$ is the value of the ramp.

Figure 4.19 shows a numerical solution to our math model using a ramp with a value $\dot{V}_{ir}$ equal to 2 volts per second for the ramp and $\Delta t = 0.05$, $\tau = 1$, and $(V_o)_{init} = 0$. Notice how the response starts off slowly and then follows the ramp input with the same slope, but with a constant time delay. Notice that the slope of the response equals the slope of the input only after around 3 seconds. This corresponds to the time it takes for the transient to die out.

The exact solution of our math model given by (4.1) to the ramp input (4.46) is found using the general method for solving differential

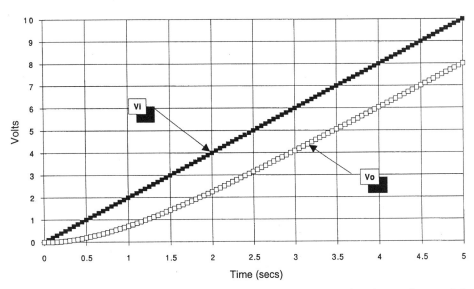

Figure 4.19. Numerical solution of equation (4.1) math model to a ramp change in input voltage
(Vr = 2 v/sec, R = 1, L = 1, Tau = RC = 1, delt = 0.05).

equations presented in the previous section titled **Exact Solution Method**. First write the solution as the sum of the homogeneous and particular solutions

$$V_o = (V_o)_H + (V_o)_P \qquad (4.47)$$

We already know that the homogeneous solution is

$$(V_o)_H = Ae^{-\frac{t}{\tau}} \qquad (4.48)$$

where the constant A must be determined using the initial condition for V_o.

You obtain the particular solution by making up a time polynomial that includes terms up to and including the ones that look like the input function. That is,

$$(V_o)_P = B_1 + B_2 t \qquad (4.49)$$

The coefficients B_1 and B_2 are obtained by substituting (4.49) into (4.1). Start with (4.1)

$$\tau \frac{dV_o}{dt} + V_o = V_i$$

and substitute (4.46) and (4.49)

$$\tau B_2 + (B_1 + B_2 t) = \dot{V}_{ir} t$$

or

$$(B_1 + \tau B_2) + B_2 t = \dot{V}_{ir} t \qquad (4.50)$$

Equation (4.50) must apply for all time so the term $(B_1 + \tau B_2)$ must equal zero and the term $B_2 t$ must equal $\dot{V}_{ir} t$ for all time. That requires

$$B_2 = \dot{V}_{ir}$$

and

$$B_1 = -\tau \dot{V}_{ir} \qquad (4.51)$$

Combining (4.47), (4.48), (4.49), and (4.51) gives

$$V_o = Ae^{-\frac{t}{\tau}} + \dot{V}_{ir}(t - \tau) \qquad (4.52)$$

We can now obtain the coefficient A from the initial condition that $(V_o)_{init} = 0$ at $t = 0$. Since the exponential is 1 when $t = 0$ we have

$$0 = A + \dot{V}_{ir}(-\tau)$$

or

$$A = \dot{V}_{ir}\tau$$

Substituting this back into (4.52) gives

$$V_o = \underbrace{\dot{V}_{ir}\tau e^{-\frac{t}{\tau}}}_{\text{transient}} + \underbrace{\dot{V}_{ir}(t - \tau)}_{\text{nontransient}} \qquad (4.53)$$

Now you can see more clearly the response of the system to a ramp input. The first term on the right side is the transient part of the solution. It gets smaller and smaller as t gets larger and larger. What remains is the nontransient second part. It shows that the output is equal to the input but displaced by the time constant τ.

Response to an Impulse Input

The most severe input a system can receive is an impulse. It's something to be avoided in the real world—a power surge is a good example. However, we do want to look at a system's response to this type of input, because in more advanced analysis of linear systems, the impulse response is commonly used as a characterization of the system. (This is primarily because it makes some of the math easier when doing high-level system analysis.)

To understand what is meant by an impulse, let's first look at the response of our system to a pulse. A pulse can be viewed as a combination of a positive step at $t = 0$ followed by a second step of the same magnitude but opposite sign at $t = t_{pulse}$. Figure 4.20 shows a numerical solution to our system math model to a pulse.

The pulse consists of a 10-volt step input beginning at $t = 0$ and an equal but opposite step at $t = 1$ second. The solution was obtained using $\Delta t = 0.05$, $\tau = 1$, and $(Vo)_{init} = 0$. Note how the response follows the previously obtained response to a step input function up to $t = 1$. At that point the solution follows what appears to be a decaying exponential.

Now let's keep the product of the step size (10 volts) and the duration (1 second) at a constant value ($10 \times 1 = 10$ volt-seconds) and reduce the pulse duration to 0.5 seconds. That means the size of the step must now equal 20 volts. Figure 4.21 shows the results. The response is obviously similar to that shown in Figure 4.20, but the output rises faster and reaches a higher voltage before the pulse ends.

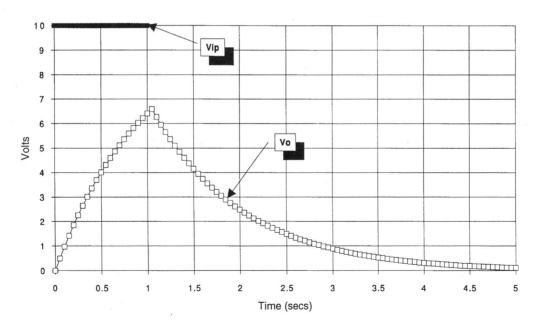

Figure 4.20. Numerical solution of equation (4.1) math model to a pulse input voltage (Vip = 10*1, R = 1, L = 1, Tau = RC = 1, delt = 0.05).

The pulse width can be further reduced until it is only the size of one time step, $\Delta t = 0.05$. The size of the step must equal 200 volts in order to produce a product equal to 10 volt-seconds. Figure 4.22 shows the response. The output imme- diately jumps to 10 volts and then slowly decays back to zero. The response shown is called the impulse response of the system. Like the step response and the ramp response, it can be used as a characteristic of the system being studied.

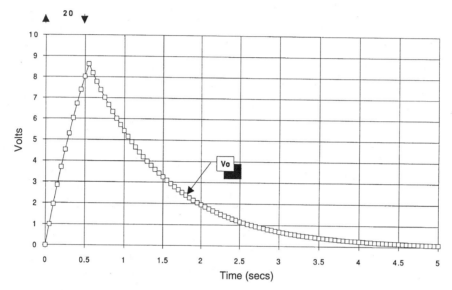

Figure 4.21. Numerical solution of equation (4.1) math model to a ramp change in input voltage (Vip = 20*0.5, R = 1, L = 1, Tau = RC = 1, delt = 0.05).

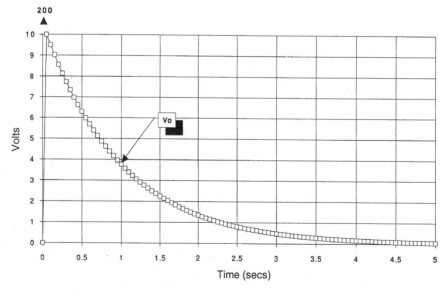

Figure 4.22. Numerical solution of equation (4.1) math model to a pulse input voltage (Vip = 200*0.05, R = 1, L = 1, Tau = RC = 1, delt = 0.05).

Now let's look at the exact solution of our system math model to an impulse function. An impulse occurs only at time $t = 0$ and then is zero for all time. Once again we write the solution as the sum of the homogeneous and particular solutions

$$V_o = (V_o)_H + (V_o)_P \qquad (4.54)$$

We know that the homogeneous solution is

$$(V_o)_H = Ae^{-\frac{t}{\tau}} \qquad (4.55)$$

where the constant A must be determined from the initial condition for V_o.

Now we run into a problem. What does the particular solution looks like for a function that only exists at time $t = 0$? The answer is that the particular solution *doesn't* exist. That is,

$$(V_o)_P = 0 \qquad (4.56)$$

So the solution to our math model for an impulse function must be

$$V_o = Ae^{-\frac{t}{\tau}} \qquad (4.57)$$

Now we are confronted with another problem. What initial condition do we use to determine the coefficient A? We know that just *prior to* $t = 0$, $V_o = 0$. We also know that at, and only at, $t = 0$ the input function takes on the

value of the impulse, which we'll label I. If we substitute this information into the math model given by (4.1) then

$$\tau \frac{dV_o}{dt} + V_o = V_i$$

becomes

$$\tau \frac{dV_o}{dt} + 0 = I$$

or

$$\left. \frac{dV_o}{dt} \right|_{t=0} = \frac{I}{\tau} \qquad (4.58)$$

This means that the proper initial condition to be used to solve for the coefficient A is given by (4.58).

Differentiating the solution given by (4.57) gives

$$\frac{dV_o}{dt} = -\frac{1}{\tau} Ae^{-\frac{t}{\tau}} \qquad (4.59)$$

Combining this with (4.58) gives

$$\frac{I}{\tau} = -\frac{1}{\tau} Ae^{-\frac{0}{\tau}} = -\frac{A}{\tau}$$

from which it is clear that A is equal to I. The response of our math model to an impulse is therefore

$$V_o = Ie^{-\frac{t}{\tau}} \qquad (4.60)$$

107

Response to an Arbitrary Input

It should be clear to you now that we can obtain the response of our system to any input which we can describe either mathematically or with a table of input data in the case of a numerical solution. Indeed, the numerical solution in the latter case may be the only solution you can obtain. Figure 4.23 shows the response of our system to such an arbitrary input function.

4.5 Power Analysis

The concepts of work, energy, and power were discussed in Chapter 2. You learned that power delivered to an element is equal to the product of the *across* and *through* variables asso-

ciated with that element. You also learned that some fundamental elements store energy while others dissipate it. We can apply these concepts to systems comprised of several basic elements. In this section, we will continue our study of the math model of our system given by equation (4.1).

Refer back to Figure 4.1 for a moment. Suppose this circuit is driven by a battery that is turned on and off by a switch (a step input). We want to know how much power this circuit will draw from the battery supplying the input voltage V_i, so we can properly size the battery. We can write the power being input to the circuit as

$$P_i = V_i \times i \qquad (4.61)$$

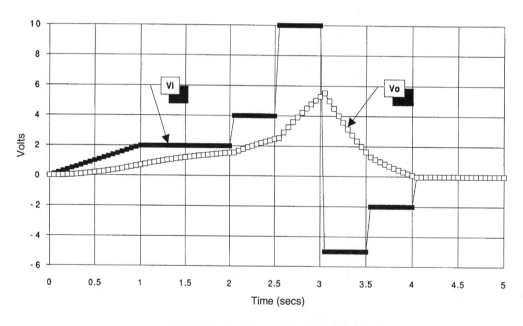

Figure 4.23. Numerical solution of equation (4.1) math model to an arbitrary input voltage (R = 1, C = 1, Tau = RC = 1, delt = 0.05).

Power Analysis for a Step Input

Let's use (4.61) to determine the power input when we apply a step input of magnitude V_{is} to the system when it has an initial condition $(V_0)_{init} = 0$. We can write the equation for the current i as

$$i = \left(\frac{V_i - V_o}{R} \right) \qquad (4.62)$$

We already solved our math model for V_0 for a step input and that solution is given in equation (4.13). Substituting (4.13) into (4.62) gives

$$i = \left(\frac{V_{is} - V_{is}\left(1 - e^{-t/\tau}\right)}{R} \right)$$

which can be simplified to

$$i = \left(\frac{V_{is}}{R} \right) e^{-t/\tau} \qquad (4.63)$$

Now we can substitute (4.63) into (4.61) to get the input power requirements. That is,

$$P_i = V_{is} \times \left(\frac{V_{is}}{R} \right) e^{-t/\tau}$$

or

$$P_i = \frac{\left(V_{is}\right)^2}{R} e^{-t/\tau} \qquad (4.64)$$

Equation (4.64) shows that the power input is an exponentially decreasing function of time.

At time $t = 0$, the power requirement is equal to

$$P_i = \frac{\left(V_{is}\right)^2}{R}$$

As time increases, the power requirement decreases and reaches zero at $t = \infty$. As you can see, the power input requirement is directly related to the current flowing into the circuit. This current, given by equation (4.63) starts off at a high value and then diminishes to zero as the voltage across the capacitor builds.

The energy delivered to the circuit is equal to the time integral of the power. That is,

$$E = \int_0^\infty \frac{\left(V_{is}\right)^2}{R} e^{-t/\tau} dt$$

or

$$E = \frac{\left(V_i^*\right)^2}{R} \left(\int_0^\infty e^{-t/\tau} dt \right) \qquad (4.65)$$

You will recall that the integral of an exponential is very easy to determine because integrals and derivatives of exponentials always end up as exponentials. In this case the integral inside the brackets in equation (4.65) is

$$\left(\int_0^\infty e^{-t/\tau} dt \right) = \tau \left[-e^{-t/\tau} \right]_0^\infty = \tau[-0 + 1] = \tau$$
$$(4.66)$$

Combining equations (4.65) and (4.66) gives the input energy as

$$E = \frac{\left(V_{is}\right)^2}{R} \tau \qquad (4.67)$$

You should take note in equations (4.67) and (4.64) how important a role the time constant has in determining the power and energy input requirements for the system. In addition, keep in mind that some of this input energy is dissipated by the resistor while some is stored in the capacitor. You can also determine how much energy was dissipated by the resistor and how much was stored in the capacitor. You know that the power loss across a resistor is equal to

$$P_R = i^2 R$$

Now substitute (4.63) for the current and get

$$P_R = \left(\frac{V_{is}}{R}\right)^2 \left(e^{-t/\tau}\right)^2 R$$

$$= \frac{\left(V_{is}\right)^2}{R} e^{-2t/\tau} \qquad (4.68)$$

The energy dissipated by the resistor is then the time integral of (4.68). That is,

$$E = \int_0^\infty \frac{\left(V_{is}\right)^2}{R} e^{-2t/\tau} dt$$

or

$$E = \frac{\left(V_{is}\right)^2}{R} \frac{\tau}{2} \qquad (4.69)$$

If you compare (4.69) with (4.67) you can see that one-half of the energy put into the system is dissipated by the resistor and the other half is stored in the capacitor.

Power Analysis for a Sinusoidal Input

We can use exactly the same procedure to compute the power delivered to our system when the input is a sinusoid. The analysis would be identical except that the solution for V_o required for equation (4.62) would be obtained from equation (4.36) or (4.45). The mathematics gets more complicated, so I am not going to take that approach right now. Instead, I want to show you how to obtain the power analysis using the numerical solution method.

Table 4.9 shows two additional columns added to my numerical solution spreadsheet for a sinusoidal input. Column E computes the input current and column F computes the input power. The numerical results are given in Table 4.10, and Figure 4.24 shows a graph of the input voltage, input current, and input power. You will note that there are times when the input power is negative. These occur when the voltage is positive and the input current is negative, or when the input voltage is negative and the input current is positive.

What meaning does negative power have? It means that instead of power being supplied to the system from the power source, the system is supplying power to the power source. The energy to do this comes from the energy stored in the system (in the capacitor) during the times when the source is supplying power to the system. You will notice that the source supplies more power to the system than it receives back from the system. This is because the resistor is dissipating energy. We'll investigate the practical effects of negative power in the next chapter.

Table 4.9.

First-order sinusoidal response with power computation.

	A	B	C	D	E	F
1	R	1				
2	C	1				
3	Tau	=B1*B2				
4	del t	=B3/20				
5	(Vo)init	0				
6	Vi*	10				
7	f	0.1				
8	Tend	=5*B3				
9						
10	t	Vi	Vo	del Vo	i	power
11	0	=$B$6*SIN(2*PI()*$B$7*A11)	=B5	=(1/$B$3*(B11-C11))*$B$4	=(B11-C11)/$B$1	=B11*E11
12	=A11+$B$4	=$B$6*SIN(2*PI()*$B$7*A12)	=C11+D11	=(1/$B$3*(B12-C12))*$B$4	=(B12-C12)/$B$1	=B12*E12
13	=A12+$B$4	=$B$6*SIN(2*PI()*$B$7*A13)	=C12+D12	=(1/$B$3*(B13-C13))*$B$4	=(B13-C13)/$B$1	=B13*E13
14	=A13+$B$4	=$B$6*SIN(2*PI()*$B$7*A14)	=C13+D13	=(1/$B$3*(B14-C14))*$B$4	=(B14-C14)/$B$1	=B14*E14
15	=A14+$B$4	=$B$6*SIN(2*PI()*$B$7*A15)	=C14+D14	=(1/$B$3*(B15-C15))*$B$4	=(B15-C15)/$B$1	=B15*E15
16	=A15+$B$4	=$B$6*SIN(2*PI()*$B$7*A16)	=C15+D15	=(1/$B$3*(B16-C16))*$B$4	=(B16-C16)/$B$1	=B16*E16
17	=A16+$B$4	=$B$6*SIN(2*PI()*$B$7*A17)	=C16+D16	=(1/$B$3*(B17-C17))*$B$4	=(B17-C17)/$B$1	=B17*E17
18	=A17+$B$4	=$B$6*SIN(2*PI()*$B$7*A18)	=C17+D17	=(1/$B$3*(B18-C18))*$B$4	=(B18-C18)/$B$1	=B18*E18
19	=A18+$B$4	=$B$6*SIN(2*PI()*$B$7*A19)	=C18+D18	=(1/$B$3*(B19-C19))*$B$4	=(B19-C19)/$B$1	=B19*E19
20	=A19+$B$4	=$B$6*SIN(2*PI()*$B$7*A20)	=C19+D19	=(1/$B$3*(B20-C20))*$B$4	=(B20-C20)/$B$1	=B20*E20
21	=A20+$B$4	=$B$6*SIN(2*PI()*$B$7*A21)	=C20+D20	=(1/$B$3*(B21-C21))*$B$4	=(B21-C21)/$B$1	=B21*E21
22	=A21+$B$4	=$B$6*SIN(2*PI()*$B$7*A22)	=C21+D21	=(1/$B$3*(B22-C22))*$B$4	=(B22-C22)/$B$1	=B22*E22
23	=A22+$B$4	=$B$6*SIN(2*PI()*$B$7*A23)	=C22+D22	=(1/$B$3*(B23-C23))*$B$4	=(B23-C23)/$B$1	=B23*E23
24	=A23+$B$4	=$B$6*SIN(2*PI()*$B$7*A24)	=C23+D23	=(1/$B$3*(B24-C24))*$B$4	=(B24-C24)/$B$1	=B24*E24
25	=A24+$B$4	=$B$6*SIN(2*PI()*$B$7*A25)	=C24+D24	=(1/$B$3*(B25-C25))*$B$4	=(B25-C25)/$B$1	=B25*E25
26	=A25+$B$4	=$B$6*SIN(2*PI()*$B$7*A26)	=C25+D25	=(1/$B$3*(B26-C26))*$B$4	=(B26-C26)/$B$1	=B26*E26
27	=A26+$B$4	=$B$6*SIN(2*PI()*$B$7*A27)	=C26+D26	=(1/$B$3*(B27-C27))*$B$4	=(B27-C27)/$B$1	=B27*E27
28	=A27+$B$4	=$B$6*SIN(2*PI()*$B$7*A28)	=C27+D27	=(1/$B$3*(B28-C28))*$B$4	=(B28-C28)/$B$1	=B28*E28
29	=A28+$B$4	=$B$6*SIN(2*PI()*$B$7*A29)	=C28+D28	=(1/$B$3*(B29-C29))*$B$4	=(B29-C29)/$B$1	=B29*E29
30	=A29+$B$4	=$B$6*SIN(2*PI()*$B$7*A30)	=C29+D29	=(1/$B$3*(B30-C30))*$B$4	=(B30-C30)/$B$1	=B30*E30
31	=A30+$B$4	=$B$6*SIN(2*PI()*$B$7*A31)	=C30+D30	=(1/$B$3*(B31-C31))*$B$4	=(B31-C31)/$B$1	=B31*E31
32	=A31+$B$4	=$B$6*SIN(2*PI()*$B$7*A32)	=C31+D31	=(1/$B$3*(B32-C32))*$B$4	=(B32-C32)/$B$1	=B32*E32
33	=A32+$B$4	=$B$6*SIN(2*PI()*$B$7*A33)	=C32+D32	=(1/$B$3*(B33-C33))*$B$4	=(B33-C33)/$B$1	=B33*E33

Table 4.10.
First-order sinusoidal response results.

	A	B	C	D	E	F
1	R	1				
2	C	1				
3	Tau	1				
4	del t	0.05				
5	(Vo)init	0				
6	Vi*	10				
7	f	0.1				
8	Tend	5				
9						
10	t	Vi	Vo	del Vo	i	power
11	0	0.00	0.00	0.00	0.00	0.00
12	0.05	0.31	0.00	0.02	0.31	0.10
13	0.10	0.63	0.02	0.03	0.61	0.38
14	0.15	0.94	0.05	0.04	0.89	0.84
15	0.20	1.25	0.09	0.06	1.16	1.46
16	0.25	1.56	0.15	0.07	1.42	2.21
17	0.30	1.87	0.22	0.08	1.65	3.10
18	0.35	2.18	0.30	0.09	1.88	4.10
19	0.40	2.49	0.40	0.10	2.09	5.20
20	0.45	2.79	0.50	0.11	2.29	6.39
21	0.50	3.09	0.62	0.12	2.47	7.65
22	0.55	3.39	0.74	0.13	2.65	8.97
23	0.60	3.68	0.87	0.14	2.81	10.34
24	0.65	3.97	1.01	0.15	2.96	11.75
25	0.70	4.26	1.16	0.15	3.10	13.19
26	0.75	4.54	1.31	0.16	3.22	14.64
27	0.80	4.82	1.48	0.17	3.34	16.10
28	0.85	5.09	1.64	0.17	3.45	17.55
29	0.90	5.36	1.82	0.18	3.54	18.98
30	0.95	5.62	1.99	0.18	3.63	20.39
31	1.00	5.88	2.17	0.19	3.70	21.77
32	1.05	6.13	2.36	0.19	3.77	23.10
33	1.10	6.37	2.55	0.19	3.83	24.39
34	1.15	6.61	2.74	0.19	3.87	25.62
35	1.20	6.85	2.93	0.20	3.91	26.78
36	1.25	7.07	3.13	0.20	3.94	27.88
37	1.30	7.29	3.33	0.20	3.96	28.90
38	1.35	7.50	3.52	0.20	3.98	29.83
39	1.40	7.71	3.72	0.20	3.98	30.69
40	1.45	7.90	3.92	0.20	3.98	31.45
41	1.50	8.09	4.12	0.20	3.97	32.11
42	1.55	8.27	4.32	0.20	3.95	32.68
43	1.60	8.44	4.52	0.20	3.93	33.15
44	1.65	8.61	4.71	0.19	3.89	33.52
45	1.70	8.76	4.91	0.19	3.86	33.78
46	1.75	8.91	5.10	0.19	3.81	33.94

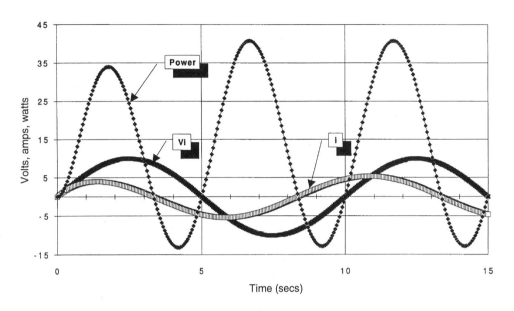

**Figure 4.24. Numerical solution of equation (4.1) math model
to a sinusoidal input voltage, showing power
(Vi = 10, f = 0.1, R = 1, C = 1, Tau = RC = 1, delt = 0.05).**

References

Press, William H. et al., *Numerical Recipes in C: The Art of Scientific Computing*, Cambridge University Press, 1988. The classic "cookbook" on numerical computation.

Sprott, Julien C., Numerical Recipes: *Routines and Examples in BASIC*, Cambridge University Press, 1991.

Chapter

5

Practical Applications of First-Order Math Models

Objectives

When you have completed this chapter, you will be able to:

- ■ **Construct a math model of a real-world engineering system using any of the four tools presented in Chapter 4.**

- ■ **Determine input power requirements for engineering systems.**

- ■ **Use the tools learned in earlier chapters to size system components, such as size a motor to a particular load.**

5.1 Introduction

In this chapter we will apply model construction and analysis techniques to a real-life design problem. The example selected involves the design of an electromechanical system that employs translational and rotational mechanical elements and an electrical motor. You will learn how to construct the mechanical and electrical circuit math models and how to use the models to analyze and design the system.

The system used in this example contains all of the fundamental characteristics of any powered drive system, regardless of the type of engineering components used. That is, the techniques shown in this example are equally applicable to power systems that utilize any combination of mechanical, electrical, and fluid components. You will learn how to use the *across* and *through* variables of each component of the system to construct power diagrams that aid in selecting and sizing components to ensure maximum efficiency and minimum size, weight, and cost.

5.2 Electromechanical System Case Study

Problem Definition

Figure 5.1 shows a sketch of a mechanical system consisting of a shaker table that runs on a pair of slider rails. We are given the task to design a controller that will move the table back and forth in a sinusoidal manner using a direct current motor. We're also given the following specifications:

Payload	*30 pounds*
Minimum period	*5 seconds*
Maximum amplitude	*1 foot*

When I attack a design problem like this, I first try to break it into steps. This particular problem can be broken down as follows:

(1) Derive the math model of the table, using any of the four methods we've discussed.

(2) Derive the math model of the motor and transmission.

(3) Derive the math model of the motor controller.

(4) Size components for steady-state loading and powering.

(5) Analyze the system's dynamic performance.

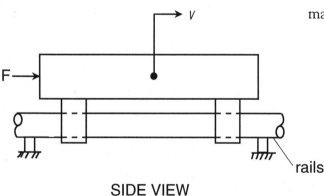

SIDE VIEW

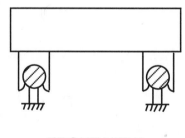

FRONT VIEW

Figure 5.1. Sketch of shaker table system.

Deriving the Math Model of the Table

We will first construct a math model of the table to gain more insight into the design problem. Initially, the table looks like it could be modeled as a pure translational mass. However, as we push other similar tables back and forth, we note that the bearings create a resistive force. The rails will be lubricated so we expect the resistive force to be proportional to velocity. We decide to model the table as a combination pure mass and pure damper.

Figure 5.2a shows our symbolic model and Figure 5.2b the associated circuit diagram. We want to find a relationship between the velocity of the table v and the applied force F. As we learned in Chapter 3, we can develop this equation using any of the following methods:

(1) Path-Vertex-Elemental Equation Method

(2) Impedance Method

(3) Operational Block Diagram Method

(4) Free-body Diagram Method.

To be on the safe side, let's use all four methods as a check to ensure that we come up with the same answer using each method.

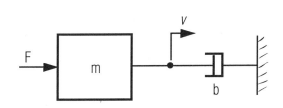

(a). *Symbolic diagram.*

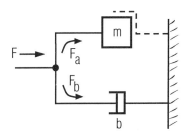

(b). *Circuit diagram.*

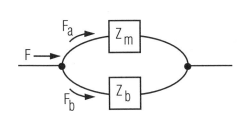

(c). *Impedance diagram.*

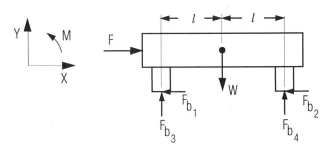

(d). *Free-body diagram.*

Figure 5.2. System diagrams.

Path-Vertex-Elemental Equation Method

From the circuit diagram we can write the elemental equations

$$F_a + F_b = F \qquad (5.1)$$

$$F_a = m\frac{dv}{dt} \qquad (5.2)$$

$$F_b = bv \qquad (5.3)$$

Substituting (5.2) and (5.3) into (5.1) gives

$$m\frac{dv}{dt} + bv = F \qquad (5.4)$$

which is the desired relationship between the velocity of the table, the applied force, the mass of the table, and the damping associated with the bearings.

Impedance Method

Again, use the circuit diagram in Figure 5.2b and replace each elemental component with its impedance as in Figure 5.2c. Since the impedances are in parallel, we write

$$\frac{v}{F} = \frac{Z_m Z_b}{Z_m + Z_b} = \frac{\dfrac{1}{mD}\dfrac{1}{b}}{\dfrac{1}{mD} + \dfrac{1}{b}} = \frac{1}{mD + b} \qquad (5.5)$$

or

$$m\frac{dv}{dt} + bv = F \qquad (5.6)$$

which is the same as equation (5.4).

Free-body Diagram Method

Figure 5.2d shows a free-body diagram of the table. Reaction forces at each slider are shown along with the weight of the table. Also shown are the positive directions for the forces and moments. Using the free-body diagram, we can write

$$\sum F_x = m\frac{dv}{dt} = F - F_{b1} - F_{b2} \qquad (5.7)$$

$$\sum F_y = 0 \quad = F_{b3} + F_{b4} - W \qquad (5.8)$$

$$\sum M = 0 \quad = F_{b4}\ell + F_{b3}\ell \qquad (5.9)$$

Equations (5.8) and (5.9) are static equations because the bearings hold the table to the rails. These equations simply tell us that F_{b3} and F_{b4} are equal and that their sum equals the weight of the table. Equation (5.7) reveals the assumption we are making regarding the resistive bearing forces. That is, we are assuming

$$F_{b1} + F_{b2} = bv \qquad (5.10)$$

Substituting equation (5.10) into (5.7) and rearranging terms gives

$$m\frac{dv}{dt} + bv = F \qquad (5.11)$$

which is the same as (5.4) and (5.6).

Block Diagram Method

Always begin the construction of a block diagram with the desired response variable as an output of an integrator placed on the right side of the diagram. We want the velocity of the table to be the response variable and the applied force to be the forcing variable. Using the elemental equations (5.1), (5.2) and (5.3), the first step is to rewrite equation (5.2) as

$$F_a = mDv$$

or

$$v = \frac{1}{mD}F_a \qquad (5.12)$$

The block diagram for equation (5.12) is sketched in Figure 5.3a.

Next rewrite equation (5.1) as

$$F_a = F - F_b \qquad (5.13)$$

The block diagram for equation (5.13) is shown in Figure 5.3b.

Finally, draw the block diagram for equation (5.3) as shown in Figure 5.3c, and connect the individual block diagrams together as shown in Figure 5.3d. This block diagram gives you quite a bit of insight into the problem. You can see that the force F applied to accelerate the mass first gets reduced by the bearing friction force bv.

When v is zero, all of the applied force goes toward accelerating the mass. As the velocity builds, less and less force goes into accelerating the mass, and more and more into overcoming the bearing friction.

The block diagram can be reduced to a single transfer function using the procedure explained in Chapter 3. That is, first write the equation for the output of the summer as

$$e = F - bv \qquad (5.14)$$

Then write the equation for v as

$$v = \frac{1}{mD}e \qquad (5.15)$$

Substitute (5.14) into (5.15) to eliminate e and rearrange the result

$$v = \frac{1}{mD}(F - bv)$$

$$mDv = F - bv$$

$$mDv + bv = F \qquad (5.16)$$

Equation (5.16) is identical to (5.4), (5.6) and (5.11).

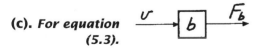

(a). For equation (5.12).

(b). For equation (5.13).

(c). For equation (5.3).

(d). Connected together.

Figure 5.3. Block diagrams.

Intuitive Solution to Table Math Model

Once I have derived a math model of a particular system, I always sit and think about it for awhile, to make sure it correlates well with the real world. I think it's important for engineers to develop a "gut" feel for what the math really means. Many problems can be spotted at this early stage by simply using your head.

All four of the methods presented in Chapter 4 have given the same results. You can decide which of the four methods you think is easier and which gives you the most insight into the problem. You can't go wrong *if you write the elemental equations correctly*. If they are correct, the only way you will come up with the wrong answer is by doing the algebra incorrectly. Always check units of the equations as you go along, to help eliminate mistakes in algebra. Let's check units now for equation (5.11).

Since force is on the right side of the equation, all terms on the left must have units of force. We know from Newton's law that mass times acceleration must have units of force. The English units of mass are lb-sec^2/ft and the units for acceleration are ft/sec^2. The product is force in pounds. We also know that velocity has units of ft/sec. The friction term b must have units of lb/ft/sec or lb-sec/ft. Divide both sides of (5.11) by b to get

$$\frac{m}{b}\frac{dv}{dt} + v = \frac{1}{b}F \qquad (5.17)$$

Then the units of m/b are

$$\left[\frac{m}{b}\right] = \frac{\text{lb} - \sec^2/\text{ft}}{\text{lb} - \sec/\text{ft}} = \sec$$

Since equation (5.17) is by now familiar, you probably already guessed that m/b is the time constant of this system. So we can rewrite (5.17) as

$$\tau\frac{dv}{dt} + v = \frac{1}{b}F \qquad (5.18)$$

where

$$\tau = m/b \qquad (5.19)$$

Compare equation (5.18) with equation (4.5) from Chapter 4. You can see that the left sides of the equations are identical. If we use the velocity–voltage (across variables) and force–current (through variables) analogy, then the right sides of the equations are also identical since F/b has units of velocity and can be thought of as an input velocity. What this means is that all the tools developed in Chapter 4 can be used to study the behavior of this mechanical math model. For example, you should be able to solve this equation numerically for $v(t)$ given $F(t)$, m, and b. You should also be able to visualize the step response for the system as shown in Figure 5.4. You should also be able to visualize the transfer function as shown in Figure 5.5a and the frequency response of the system as shown in Figure 5.5b. In short, the math model of this mechanical system and that of the electrical system discussed so extensively in Chapter 4 are identical. So we know that this mechanical circuit will act like a low-pass filter (mechanical style filter). The table in essence will move with a velocity equal to F/b when a low-frequency sinusoidal force is applied. As the frequency of the force increases past the break frequency ($f = 1/2\pi\tau$), the velocity will become smaller than F/b.

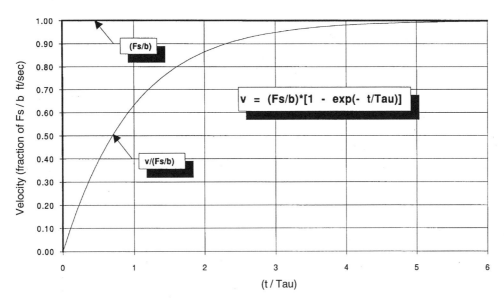

Figure 5.4. Response of equation (5.18) to a step change in F.

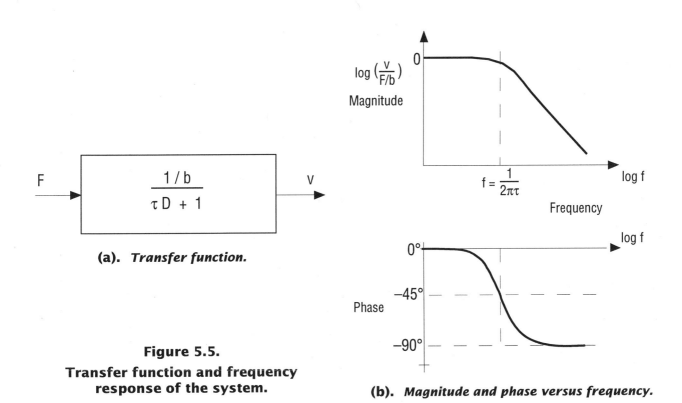

(a). *Transfer function.*

Figure 5.5.
Transfer function and frequency response of the system.

(b). *Magnitude and phase versus frequency.*

Power Requirements

Let's now turn our attention to a very important matter in designing any powered engineering system—the load or input power requirement. We are going to drive the table with a DC motor and we need to determine what power rating the motor must have. According to the specifications given to us, the table is to move back and forth in a sinusoidal motion with a maximum displacement of 1 ft. Let's call this maximum displacement X_o and write the displacement at any point in time as

$$x(t) = X_o \sin 2\pi ft \qquad (5.20)$$

We want to determine the power requirements as a function of the frequency of the sinusoidal motion. We can obtain the velocity of the table by differentiating equation (5.20)

$$v = \dot{x} = 2\pi f X_o \cos 2\pi ft \qquad (5.21)$$

and the acceleration by differentiating once more

$$\dot{v} = \ddot{x} = -(2\pi f)^2 X_o \sin 2\pi ft \qquad (5.22)$$

Now we can substitute (5.21) and (5.22) into (5.11) to obtain the force required to produce the desired sinusoidal motion.

$$F(t) = m\left[-(2\pi f)^2 X_o \sin 2\pi ft\right] + b\left[2\pi f X_o \cos 2\pi ft\right] \qquad (5.23)$$

Note that this substitution is the reverse of the problem, "Given $F(t)$, find $v(t)$." When we are given information that allows us to determine

$v(t)$, there is no need to solve the differential equation. Yet, the differential equation still applies and is very useful.

I've placed $F(t)$ on the left side of the equation to indicate that it is the dependent variable. The power requirement, $F(t) \times v(t)$, can be obtained by multiplying equations (5.21) and (5.23). That is

$$Pow(t) = F(t) \times v(t)$$
$$= m\left[-(2\pi f)^3 X_o^2 \sin 2\pi ft \cos 2\pi ft\right]$$
$$+ b\left[(2\pi f X_o \cos 2\pi ft)^2\right] \qquad (5.24)$$

You can see from this equation that the power required is composed of two terms. One is associated with and dependent on the table damping b and the other with the table mass m.

It is difficult to visualize the power requirements just by looking at (5.24). You can see it better by plotting (5.24) as a function of time with the frequency held constant. Of course, to do this you must select values for m, b, and f.

To select these values we return to the specification for guidance. The table is supposed to carry 30 lbs. Let's say we estimate at this stage of the design that the table will weigh 2.2 lbs. That gives a total weight W equal to 32.2 lbs or a mass of

$$m = W/g = \frac{32.2 \text{ lbs}}{32.2 \text{ ft/sec}^2} = 1 \frac{\text{lb} - \text{sec}^2}{\text{ft}}$$

A value for the damping b is not given in the specification. We estimate it from a brief lab test on a similar table at work to be equal to 1 lb-sec/ft. We're told in the specification that the minimum period of the table is 5 sec. That's a frequency of 0.2 Hz. We note that the break frequency with a value of b equal to 1 lb-sec/ft would be

$$f = \frac{1}{2\pi\tau} = \frac{b}{2\pi m} = 0.159\,\text{Hz}$$

That means the desired frequency is going to be past the break frequency. We also note that the time constant of the table will be $\tau = m/b = 1/1 = 1$ sec.

I set equation (5.24) up on a spreadsheet and evaluated $Pow(t)$ over one complete cycle using $m = 1$ lb-sec^2/ft, $b = 1$ lb-sec/ft, $X_o = 1$ ft, and $f = 0.2$ Hz. A plot of my results is shown as Figure 5.6a. Study these results carefully. You can see that the total power fluctuates and has two positive and two negative peaks during one cycle. To help understand this, I also plotted the power component due to the mass and the damper. The mass, of course is an energy storage device and hence absorbs and then gives back power. The damper, on the other hand, always consumes power.

Look at Figure 5.6b where I have plotted $x(t)$, $v(t)$, $a(t)$ and the product $a(t)*v(t)$. The latter product is directly proportional to the power associated with the mass as indicated by equation (5.24). You can clearly see that it is this product that gives rise to the two positive and two negative power peaks.

Now look at Figure 5.7a. This shows the force required to produce the desired $x(t)$ and the resulting velocity plotted against time for one cycle. The product of these two variables, as you know, is power. You can clearly see that there are two regions where the product of the two is negative. Around $t = 1.25$ to $t = 2.00$ the force is positive and the velocity is negative, resulting in negative power. Around $t = 3.75$ to $t = 4.5$, the velocity is positive and the force is negative, again resulting in negative power. During these two time intervals, power will be returned to the DC motor that we plan to use to drive the table.

Another very useful way to look at the power requirements of a system is to plot F against v. To do this, select sequential points in time in Figure 5.7a and find F and v. Plot these pairs as a single point in an F versus v plane as shown in Figure 5.7b. Such plots are often called *load curves* or *load characteristics*. They describe a load in such a way that they can easily be superimposed on top of motor performance data, or so-called *drive characteristics*. You will often see data on electric motors, hydraulic motors, and prime movers (gasoline and diesel engines, for example) plotted with their *through variable* (force, torque, current, flow, etc.) on one axis and their *across variable* (linear speed, angular speed, voltage, pressure, etc.) on the other. By superimposing these *drive characteristics* on top of the *load characteristics*, you can determine if the motor will be able to drive the load.

The load characteristic shown in Figure 5.7b is typical of a load that contains a so-called *reactive* element—that is, an energy storage element. If the load were purely resistive in nature, the force would be in phase with the velocity as shown in Figure 5.8a, the load curve would look like that shown in Figure 5.8b, and

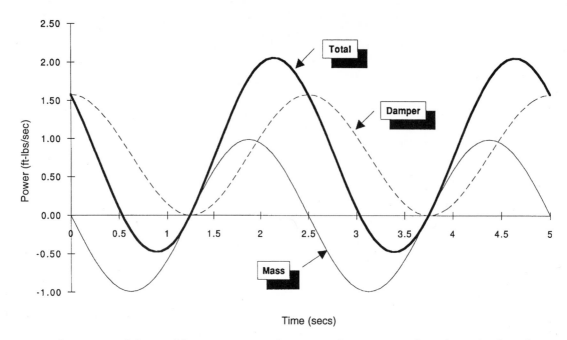

Figure 5.6(a). Table power requirement from equation (5.24) showing components due to mass and damper (f = 0.2 Hz).

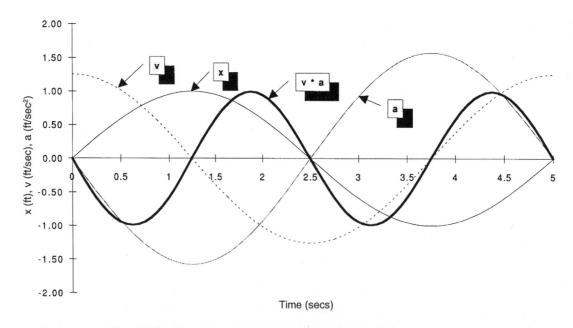

Figure 5.6(b). Acceleration, velocity, and displacement of table (f = 0.2 Hz).

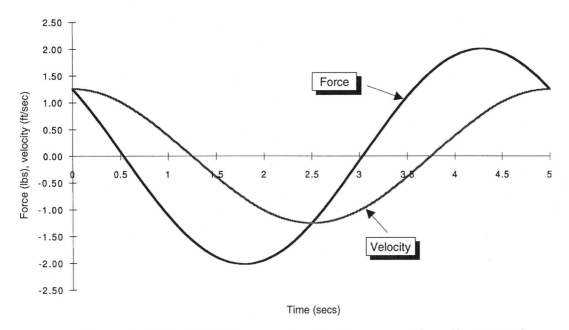

Figure 5.7(a). Table force and velocity versus time (f = 0.2 Hz).

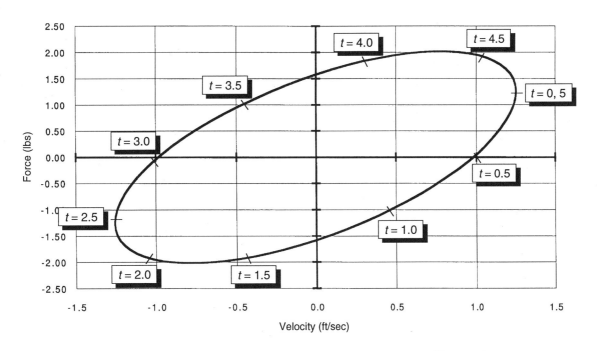

Figure 5.7(b). Table force versus velocity (f = 0.2).

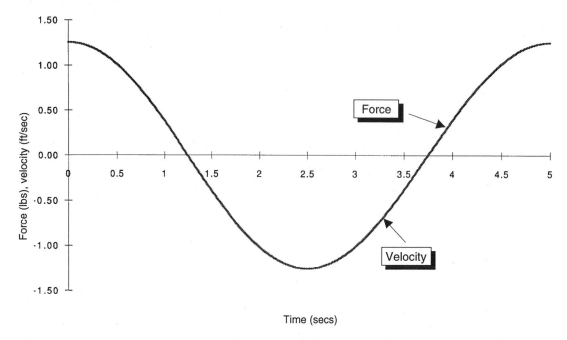

Figure 5.8(a). Table force and velocity versus time for purely resistive load (f = 0.2 Hz).

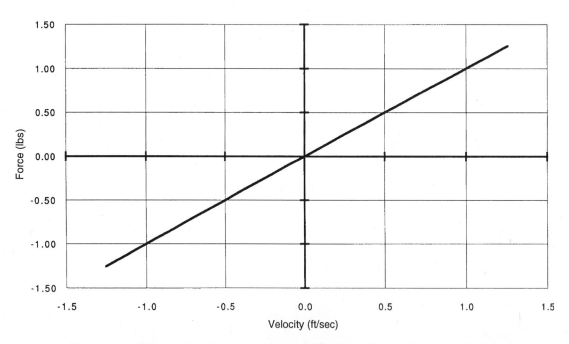

Figure 5.8(b). Table force versus velocity for purely resistive load (f = 0.2 Hz).

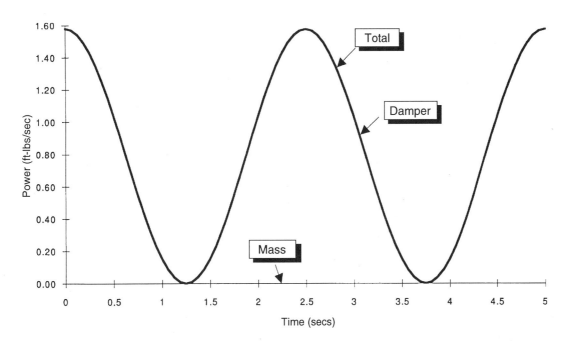

Figure 5.8(c). Table power requirement for purely resistive load (f = 0.2 Hz).

the power as a function of time would look like Figure 5.8c. This latter figure shows that, while the power varies from zero to a peak value, there is an average power that must be input to the system that is then dissipated by the resistive load.

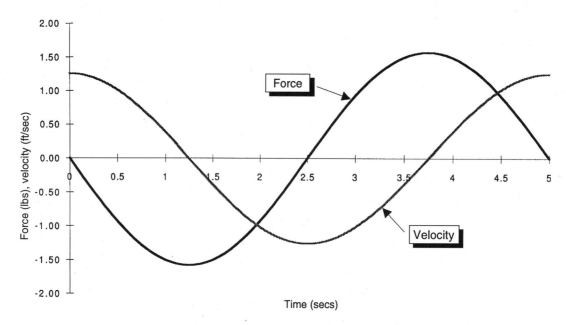

Figure 5.9(a). Table force and velocity versus time for pure mass load (f = 0.2 Hz).

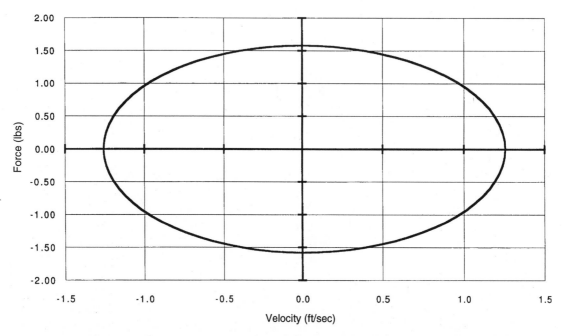

Figure 5.9(b). Table force versus velocity for pure mass load (f = 0.2 Hz).

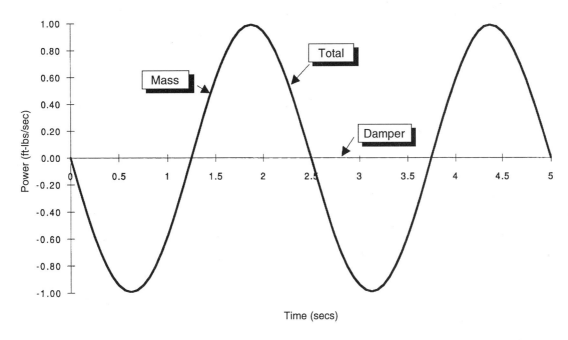

**Figure 5.9(c). Table power requirement for pure mass load
(f = 0.2 Hz).**

On the other hand, if the load were purely reactive (that is, it contained no energy dissipative elements) then the force and velocity would be 90 degrees out of phase as shown in Figure 5.9a, the load curve would look like that shown in Figure 5.9b, and the power as a function of time would look like Figure 5.9c. This latter figure shows that, on average, no power is required. Real-life systems fall somewhere between these two extremes.

So far, we have looked at only one table frequency. However, equation (5.24) clearly shows that the power requirements are a function of frequency. For comparative purposes I ran two cases where everything was held constant but the frequency was halved (f = 0.1 Hz) and doubled (f = 0.4 Hz). The results, along with the results for the specification frequency of f = 0.2 Hz, are shown in Figure 5.10. Study these plots carefully. You will note that lowering the frequency caused the power requirements to decrease significantly and made the load look more like a pure resistive load. On the other hand, increasing the frequency drastically increased the power requirement and made the load appear more like a purely reactive load.

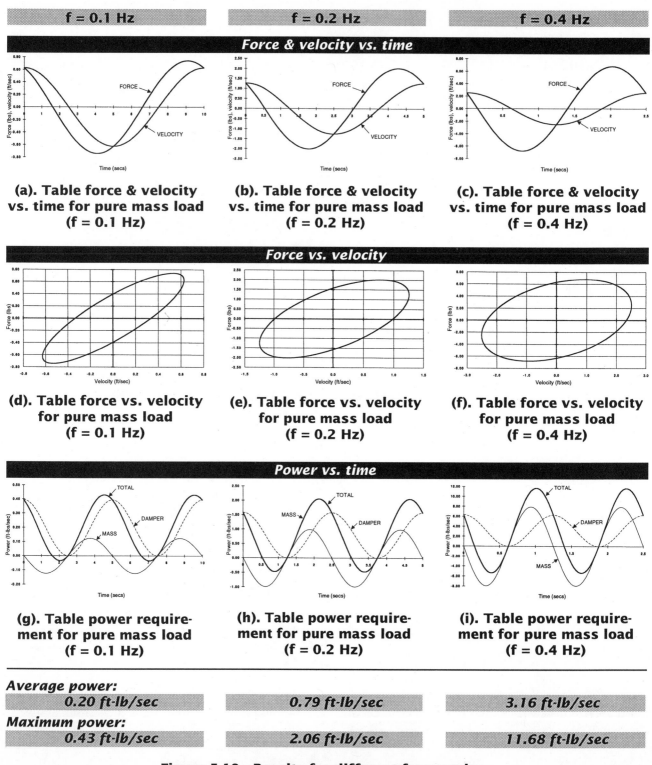

Figure 5.10. Results for different frequencies.

Derivation of Math Model of Motor and Transmission

We will now continue our study of the shaker table and look into the modeling and sizing of the DC motor and the transmission that will connect the motor to the shaker table. Figure 5.11 shows a sketch of one of many ways in which a DC motor might be used to drive the shaker table. The table is attached at two ends with a taut, inelastic cable wound around an idler pulley at one end and a drive pulley at the other.

If the pulley is not accelerating then the relationship between torque, Q, exerted on the pulley by the motor and the force F exerted on the table and in the cable is

$$Q = rF \qquad (5.26)$$

Equations (5.25) and (5.26) are fundamental transmission equations. They provide a mathematical relationship to convert angular speed into linear speed and vice versa, and to convert linear force into rotational torque and vice versa.

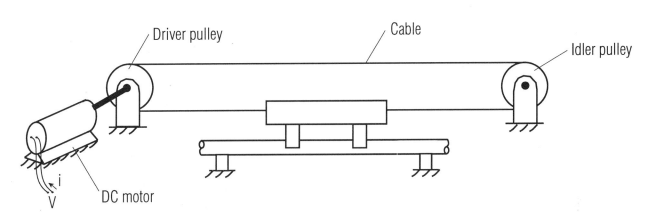

Figure 5.11. Sketch showing DC motor driving shaker table.

You can see that the cable and pulley act as a form of mechanical transformer or transmission. They convert translational motions into rotary motions and translational forces into torques, and vice versa. Figure 5.12 shows the relationships. The linear velocity of any point at a distance r from the center of rotation of a pulley is given by

$$v = r\omega \qquad (5.25)$$

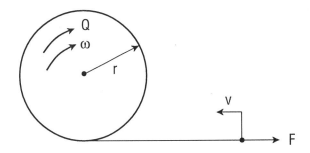

Figure 5.12. Relationship between translational motions and forces and rotary motions and torques.

Figure 5.13 shows a free-body diagram model of the motor armature and pulley. The armature of the motor has a moment of inertia J_m ft-lb-sec^2 and we will assume that it is much larger than the moment of inertia associated with the pulleys. We will also assume that the bearings in the motor and pulley act as a rotational damper having a damping coefficient, B ft-lb/rad/sec. Current flowing through the motor armature creates a motor torque Q_m ft-lbs. This torque is resisted by the bearing damper torque Q_B ft-lbs, and the load placed on the motor output shaft Q_L ft-lbs. From the free-body diagram we can write,

$$Q_m - Q_B - Q_L = J_m \frac{d\omega}{dt} \qquad (5.27)$$

where

$$Q_B = B\omega \qquad (5.28)$$

Combining equations (5.27) and (5.28) and then rearranging gives,

$$J_m \frac{d\omega}{dt} + B\omega = Q_m - Q_L \qquad (5.29)$$

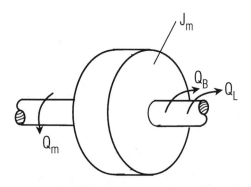

Figure 5.13. Free-body diagram of motor armature and pulley.

The load torque Q_L can be related to the linear load force F_L exerted on the table using equation (5.26). That is,

$$Q_L = rF_L \qquad (5.30)$$

Similarly the linear drive force created by the motor can be expressed as

$$F_m = \frac{Q_m}{r} \qquad (5.31)$$

Note that we can use equation (5.25) to change the variable ω in (5.29) to a linear velocity. Let's do this and then substitute equations (5.30) and (5.31) into the result. Start with (5.29)

$$J_m \frac{d\omega}{dt} + B\omega = Q_m - Q_L$$

Substitute (5.25)

$$\frac{J_m}{r}\frac{dv}{dt} + \frac{B}{r}v = Q_m - Q_L$$

Substitute (5.31)

$$\frac{J_m}{r}\frac{dv}{dt} + \frac{B}{r}v = rF_m - rF_L$$

and rearrange terms

$$\frac{J_m}{r^2}\frac{dv}{dt} + \frac{B}{r^2}v = F_m - F_L \qquad (5.32)$$

Now let's substitute the table model as expressed by equation (5.11) for F_L,

$$\frac{J_m}{r^2}\frac{dv}{dt} + \frac{B}{r^2}v = F_m - m\frac{dv}{dt} - bv$$

and rearrange to give

$$\left(\frac{J_m}{r^2} + m\right)\frac{dv}{dt} + \left(\frac{B}{r^2} + b\right)v = F_m \quad (5.33)$$

We have taken several big and very important steps, so I want to pause for a moment and review them with you. First, we wrote the elemental equations of the motor showing the torque developed by the motor and the load torque as independent variables. That is,

$$J_m\frac{d\omega}{dt} + B\omega = Q_m - Q_L$$

This is a mechanical math model of the motor. You should recognize it as a first-order differential equation and readily spot that the time constant is given by

$$\tau_m = \frac{J_m}{B}$$

Then I introduced the simple concept of a transmission, showing that linear force and velocity are related to rotational torque and angular speed by the radius of a pulley. I then used these transmission relationships to convert the rotational math model of the motor into the translational equivalent

$$\frac{J_m}{r^2}\frac{dv}{dt} + \frac{B}{r^2}v = F_m - F_L$$

Finally I took the translational mechanical model we developed for the table and treated it as the load that the motor will see. That is, I

rewrote equation (5.11) as

$$F = F_L = m\frac{dv}{dt} + bv$$

and substituted this into the motor equation to give the final result

$$\left(\frac{J_m}{r^2} + m\right)\frac{dv}{dt} + \left(\frac{B}{r^2} + b\right)v = F_m$$

Note that the equation is still a first-order ordinary linear differential equation with constant coefficients. It looks like equation (5.11) but now the coefficients have a term from the rotational equation divided by the radius of the pulley.

Let's check the units just to make sure the equation is correct. First, we can see that the units of J_m / r^2 must equal those of m. That is,

$$\left[\frac{J_m}{r^2}\right] = \frac{\text{ft} - \text{lb} - \sec^2}{\text{ft}^2} = \frac{\text{lb} - \sec^2}{\text{ft}}$$

$$[m] = \frac{F}{a} = \frac{\text{lb} - \sec^2}{\text{ft}}$$

so these check. Next, the units of B/r^2 must equal those of b. That is,

$$\left[\frac{B}{r^2}\right] = \text{ft} - \text{lb} - \sec \times \frac{1}{\text{ft}^2} = \frac{\text{lb} - \sec}{\text{ft}}$$

$$[b] = \frac{F}{v} = \frac{\text{lb} - \sec}{\text{ft}}$$

So these also check. We have checked units before for the rest of the equation, so the units are all fine. You should also note that the new time constant of the motor-table system is given by

$$\tau_{mT} = \frac{\dfrac{J_m}{r^2} + m}{\dfrac{B}{r^2} + b} = \frac{J_m + r^2 m}{B + r^2 b} \quad (5.34)$$

The transmission equations given in (5.25) and (5.26) can also be used to describe the motor-table system in terms of torque and angular velocity. To do this, we begin again with equation (5.29) but this time we leave the left side alone and substitute (5.11) into (5.30) and then that result into (5.29), giving

$$J_m \frac{d\omega}{dt} + B\omega = Q_m - r\left[m\frac{dv}{dt} + bv \right] \quad (5.35)$$

Then using (5.25) to convert linear to angular velocity, we get

$$J_m \frac{d\omega}{dt} + B\omega = Q_m - r^2 m \frac{d\omega}{dt} - r^2 b\omega \quad (5.36)$$

Collecting like terms gives

$$\left(J_m + r^2 m\right)\frac{d\omega}{dt} + \left(B + r^2 b\right)\omega = Q_m \quad (5.37)$$

or in terms of the time constant

$$\tau_{mT}\frac{d\omega}{dt} + \omega = \frac{1}{B + r^2 b} Q_m \quad (5.38)$$

where you can see that the time constant is the same as that given by (5.34).

A block diagram of the motor-table system as described by equation (5.38) is given in Figure 5.14. We will use this block diagram to connect to the electrical portion of the motor model next.

Derivation of Math Model of Motor Controller

A direct current permanent magnet motor (DC-PMM) is proposed to control the desired sinusoidal motion of the table. Figure 5.15 shows a circuit schematic that can be used to describe a DC-PMM. The magnetic field of such a motor is created by a permanent magnet. The voltage E_c applied across the armature will be resisted by the armature resistance R_a, the armature inductance L_a, and the back electromotive force E_{bemf} induced by the armature winding rotating in the magnetic field.

The torque developed by a DC motor is proportional to the armature current. That is,

$$Q_m = K_1 i_a$$

where K_1 = a motor constant. $\qquad$ (5.39)

The voltage generated by a motor is proportional to its speed. This is called back voltage and is given by

$$E_{bemf} = K_2 \omega \qquad (5.40)$$

where K_2 = another motor constant
and ω = the speed of the motor.

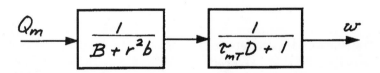

Figure 5.14. Block diagram of motor–table system.

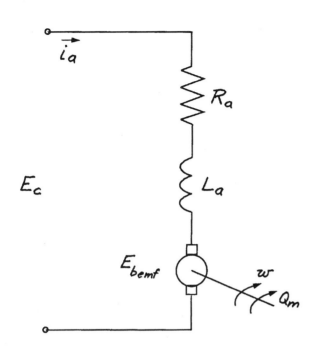

Figure 5.15. Circuit schematic used to describe a DC-PMM.

From the circuit diagram given in Figure 5.15, we can write the voltage around the loop as

$$E_c - R_a i_a - L_a \frac{di_a}{dt} - E_{bemf} = 0 \quad (5.41)$$

Substituting (5.40) into (5.41) and rearranging gives

$$L\frac{di_a}{dt} + R_a i_a = E_c - K_2 \omega \quad (5.42)$$

You will recognize (5.42) as a first-order ordinary linear differential equation with constant coefficients relating the armature current as a function of the input control voltage and the back voltage of the motor. We can also write this equation in terms of an armature circuit time constant

$$\tau_a \frac{di_a}{dt} + i_a = \frac{1}{R_a}\left(E_c - K_2\omega\right) \quad (5.43)$$

or

$$\left(\tau_a D + 1\right)i_a = \frac{1}{R_a}\left(E_c - K_2\omega\right) \quad (5.44)$$

where

$$\tau_a = \frac{L_a}{R_a} \quad (5.45)$$

A block diagram of the motor as represented by equations (5.44) and (5.39) is shown in Figure 5.16. You can easily see how this block diagram will fit together with the mechanical motor-table model given in Figure 5.14.

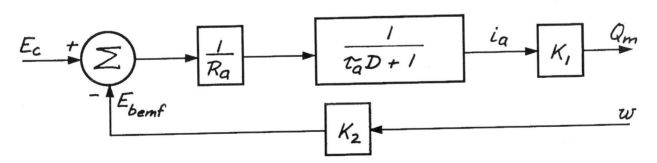

Figure 5.16. Block diagram of motor.

In steady state, di_a / dt in equation (5.43) is zero. That is,

$$i_a = \frac{E_c - K_2 \omega}{R_a} \qquad (5.46)$$

We can substitute equation (5.46) into (5.39) to obtain the steady-state characteristics of the motor. That is,

$$Q_m = K_1 i_a = \frac{K_1}{R_a}\left(E_c - K_2 \omega\right)$$

or

$$Q_m = \frac{K_1}{R_a} E_c - \frac{K_1 K_2}{R_a} \omega \qquad (5.47)$$

Since K_1, K_2, and R_a are constants, you can see that for a given applied motor voltage E_c, (5.47) is the equation for a straight line. These motor characteristics are shown in Figure 5.17 where a family of lines has been created for several values of E_c.

The power characteristics of this motor can be obtained by multiplying the torque of the motor given by equation (5.47) by the angular velocity, ω. That is

$$Pow_M = Q_M \omega$$

$$= \frac{K_1}{R_a} E_c \omega - \frac{K_1 K_2}{R_a} \omega^2 \qquad (5.48)$$

By setting E_c to its maximum allowable value, we can obtain the power versus speed curve shown in Figure 5.18.

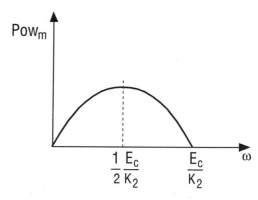

Figure 5.18. Power versus speed curve.

You can see from this plot that the motor can only develop maximum power at one speed. That speed can be found by differentiating equation (5.48), setting the result to zero and solving for ω. The result is

$$\omega_{\text{max power}} = \frac{1}{2}\frac{E_c}{K_2} \qquad (5.49)$$

The maximum power speed is therefore one-half the so-called *free running* speed of the motor. The free running speed is the speed when there is no load on the motor.

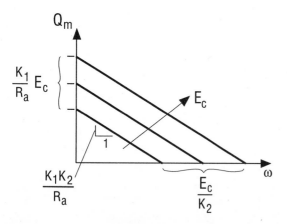

Figure 5.17. Steady-state characteristics of the motor.

Substituting (5.49) into (5.48) gives the maximum power

$$Pow_{M\,max} = \frac{1}{4}\frac{K_1}{K_2 R_a}E_c^2$$

$$= \frac{1}{4}\frac{K_1}{K_2}E_c i_a \qquad (5.50)$$

Since E_c times i_a is the electrical input power to the motor, the efficiency of the motor at its maximum power point can be obtained by dividing (5.50) by $E_c i_a$. That is,

$$\eta_M = \frac{Pow_{M\,max}}{i_a E_c} = \frac{1}{4}\frac{K_1}{K_2} \qquad (5.51)$$

Steady-state Design

Before you analyze the dynamics of a system you're designing, you must size components to handle the steady or quasi-steady loads and powering requirements. Then you can perform a dynamic analysis in which you subject the system to severe loadings, such as step responses, to determine if the design, based on steady-state loads, can handle the dynamic loads.

In this case study, we are designing a system that will experience continuous sinusoidal motions and loads. As we saw above, the power requirements are cyclic, but repeatable. A system in this state is sometimes referred to as being in a *quasi-steady* state.

We now know the motor characteristics so we can use the power analysis techniques developed earlier to determine the size of the motor.

The approach will be to use equation (5.37), describing the torque requirements of the mechanical system, to reflect these loads onto the motor characteristics.

The displacement, velocity, and acceleration of the shaker table were previously given by equations (5.20), (5.21), and (5.22), respectively. We can use the transmission equation (5.25) to convert these to angular rotation. That is,

$$\theta = \frac{X_o}{r}\sin 2\pi f t \qquad (5.52)$$

$$\omega = 2\pi f \frac{X_o}{r}\cos 2\pi f t \qquad (5.53)$$

$$\dot\omega = -(2\pi f)^2 \frac{X_o}{r}\sin 2\pi f t \qquad (5.54)$$

Now we can rearrange (5.37) so Q_m is a dependent variable and then substitute (5.53) and (5.54) for ω and $d\omega / dt$ respectively.

$$Q_{mL} = (J_M + r^2 m)\left[-(2\pi f)^2 \frac{X_o}{r}\sin 2\pi f t\right]$$

$$+ (B + r^2 b)\left[2\pi f \frac{X_o}{r}\cos 2\pi f t\right] \qquad (5.55)$$

I've designated this as the Q_{mL} to indicate it is the *load* torque requirement, not the torque *developed* by the motor. In steady state the two are equal. During transients they are not.

Now let's use (5.55) and (5.53) to prepare a plot of Q_{mL} versus ω. Then we can superimpose this load requirement on the steady motor Q_m versus ω characteristics. From our early design work and the system specifications we have:

$$m = 1 \frac{\text{lb} - \text{sec}^2}{\text{ft}}$$

$$b = 1 \frac{\text{lb} - \text{sec}}{\text{ft}}$$

$$X_o = 1 \, \text{ft}$$

$$f = 0.2 \, \text{Hz}$$

We still need the radius of the driver pulley, the motor inertia and the motor damping. The motor inertia is generally given by the motor manufacturer, but not always. Motor damping is generally not given. You may have to conduct experiments to determine these values or estimate them from past experience. To complicate matters even more, the motor inertia will depend on the size of the motor chosen, which is what we're trying to determine. When you are faced with such "Catch 22" situations, you must make an educated guess and then proceed to determine if the guess was correct. If not, you make a better guess and continue to do this until the design converges.

Say we select a DC-PMM with the following characteristics:

J_M = 1 ft-lb-sec^2

B = 1 ft-lb-sec

E_c = 12 volts

R_a = 1.83 ohms (volts/amp)

K_1 = 3.210 ft-lb/amp

K_2 = 1.146 volt-secs

N_{FR} = 100 RPM = free running speed of motor

We know that the maximum power speed of a DC-PMM is about one-half the free-running speed. To size the pulley, we will initially guess from our previous design work that the maximum power load requirement will occur at about 0.8 times the maximum table velocity. That is

$$v_{\text{max power}} = 0.8 \times 2\pi f X_o$$

$$= 0.8 \times 2\pi \times 0.2 \times 1$$

$$= 1.0 \, \text{ft / sec}$$

We can now select the radius of the pulley using

$$r = \frac{v_{\text{max power}}}{\omega_{\text{max power}}}$$

$$= \frac{1}{\frac{2\pi(50)}{60}}$$

$$= \frac{60}{100\pi}$$

$$= 0.191 \, \text{ft (2.29 in)}$$

This is an odd size pulley, so select the nearest size of 2.5 inches (0.208 ft).

Take a moment now and notice how the transmission equation was used to place the speed at which the maximum table load occurs, at the motor speed where the motor develops maximum power. This ensures that the motor operates at maximum efficiency, thereby reducing the size and cost of the motor. Mechanical transmissions are often used for this purpose, just as transformers are used for this purpose in electrical systems.

We now have enough information to solve equation (5.55) for load torque and (5.53) for load angular speed. I solved these two equations for one complete cycle using a spreadsheet. My results are plotted in Figure 5.19 along with the motor torque-speed characteristics for $E_c = \pm 12$ volts. You can see that the load characteristic is slightly below the maximum motor characteristic, so the design is satisfactory. Keep in mind that Figure 5.19 shows only the maximum motor torque versus speed characteristics. When the system is operating, the applied motor voltage E_c is varied so that the torque developed by the motor is approximately equal to the load torque at any instant in time.

Dynamic Performance Analysis

Now that you understand how to size a motor to a load, let's combine the electrical model with the mechanical model and investigate the dynamic performance of the system. The block diagram of the combination is given in Figure 5.20.

This diagram is simply Figures 5.14 and 5.16 pieced together. Take some time now to study this model. You can see that if the system were started from rest (E_c and $\omega = 0$), then a step change in E_c would not produce an immediate motor torque Q_m due to the time constant associated with the resistance and inductance of the motor armature. After a while, the current starts to flow and the motor starts developing torque. This torque does not instantaneously change

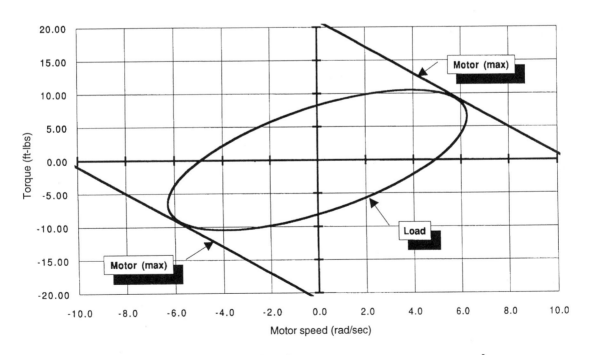

Figure 5.19. Motor and load torque versus motor speed
(f = 0.2 Hz, r = 0.20 ft).

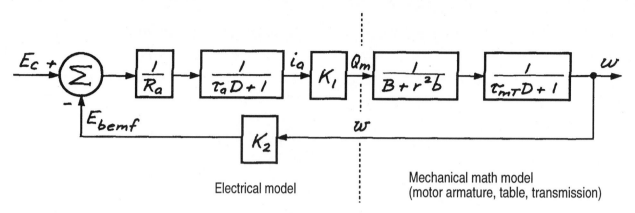

Figure 5.20. Block diagram of the total system.

the speed of the motor, because of the inertia associated with the motor armature and table. A moment later the motor does respond and starts creating a back voltage, which reduces the voltage applied to the armature. This continues until the voltage applied to the motor is reduced enough so that the current flowing through the armature generates just the right amount of torque to overcome the load.

The combined electromechanical model of our system is described by two first-order ordinary linear differential equations. The output of one is the input to the other. Let's look at these two equations again. Equation (5.43) repeated below

$$\tau_a \frac{di_a}{dt} + i_a = \frac{1}{R_a}\left(E_c - K_2 \omega\right) \qquad \begin{matrix}(5.43)\\ \text{repeated}\end{matrix}$$

governs the current flowing through the armature of the motor given an applied voltage and back voltage proportional to the speed of the motor. Equation (5.39) repeated below

$$Q_m = K_1 i_a \qquad \begin{matrix}(5.39)\\ \text{repeated}\end{matrix}$$

converts the armature current into the torque developed by the motor.

Equation (5.39) can be substituted into (5.43) to convert it into a first-order differential equation for motor torque. That is,

$$\tau_a \frac{dQ_m}{dt} + Q_m = \frac{K_1}{R_a}\left(E_c - K_2 \omega\right) \qquad (5.56)$$

Next we have the first-order differential equation (5.38) that governs the motor speed, given the motor torque. That is,

$$\tau_{mT} \frac{d\omega}{dt} + \omega = \frac{1}{B + r^2 b} Q_m \qquad \begin{matrix}(5.38)\\ \text{repeated}\end{matrix}$$

As you can see, equations (5.56) and (5.38) are two simultaneous first-order linear ordinary differential equations with constant coefficients.

Solving two first-order differential equations simultaneously is no more difficult than solving one. You can solve them numerically exactly as you did before. That is, you rewrite the two equations so only the derivative is on the left side of the equation and then you use the delta approximation. For these two equations we have

$$\Delta Q_m = \frac{1}{\tau_a}\left[\frac{K_1}{R_a}(E_c - K_2\omega) - Q_m\right]\Delta t \quad (5.57)$$

and

$$\Delta\omega = \frac{1}{\tau_{mT}}\left[\frac{1}{(B + r^2 b)}Q_m - \omega\right]\Delta t \quad (5.58)$$

We start the solution with two initial values instead of one

$$(Q_m)_{init} = Q_{mo}$$

$$(\omega)_{init} = \omega_o$$

Then given $E_c(t)$, we solve (5.57) and (5.58) for the incremental changes in Q_m and ω. These are

added to the old values of Q_m and ω, time is incremented, and the process repeated.

I first solved (5.57) and (5.58) for a step change in E_c from 0 to 12 volts with the two initial conditions at zero. I used a time step equal to $\tau_a / 10$ since this is the *smaller* of the two time constants. My results are shown in Figure 5.21.

Notice in this figure how the motor torque jumps to almost twice its steady-state value. Since motor torque is proportional to armature current [see equation (5.39)], you can see that the motor initially draws almost twice the current from the voltage (power) source than it does when the motor is running in steady state. You will recall the power dissipated in a resistor is equal to the current times the square of the resistor. In the case of the motor, the resistor is the armature windings, so we would want to

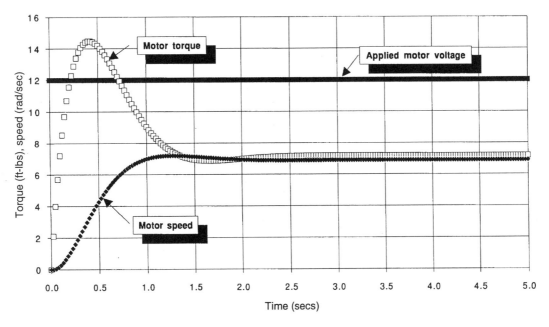

Figure 5.21. Numerical solution of motor–table math model to step motor voltage input [see equations (5.57) and (5.58)].

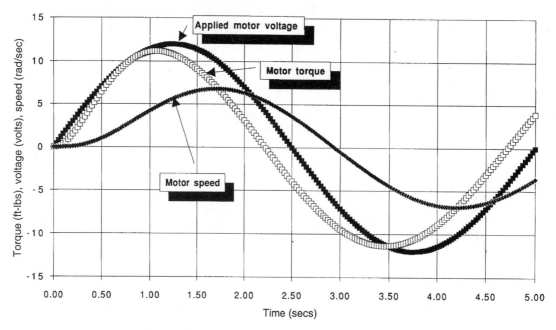

Figure 5.22. Numerical solution of motor–table math model to sinusoidal motor voltage input [see equations (5.57) and (5.58)].

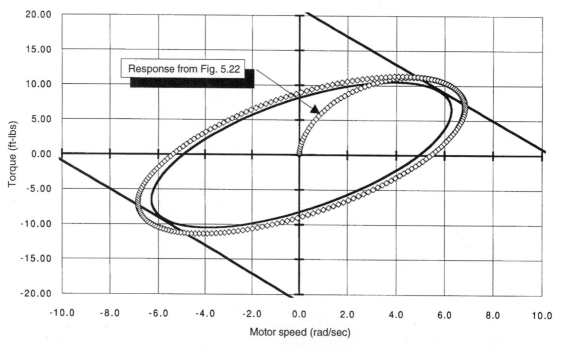

Figure 5.23. Sinusoidal response plotted on motor/load torque versus speed characteristics.

check to make sure the armature windings could temporarily withstand this much current. Also note that the speed of the motor starts off slowly and then builds up rapidly toward its steady-state value. The motor actually overshoots the steady-state speed slightly.

I also solved (5.57) and (5.58) for a suddenly applied sinusoidal input voltage E_c varying between 12 and –12 volts with the two initial conditions at zero. I used the same time step. My results are shown in Figure 5.22.

A sinusoid is a much gentler input to a system than is a step. In the beginning, a sinusoid looks almost like a ramp input. You can see that the peak motor torque is considerably less than it was for the step input. Notice, too, how the motor torque leads the input voltage as the system approaches quasi-steady state. That's because the motor current is leading the input voltage—a phenomena we will explore more in the next chapter.

I took the motor torque and motor speed results from Figure 5.22 and plotted them on the quasi-steady state motor and load characteristic that we developed earlier (Figure 5.19). My results are shown in Figure 5.23. This shows a little more clearly how the elliptical load pattern gets created as the system approaches the quasi-steady state condition. Notice also that the dynamic solution is indicating that the motor is being required to operate a portion of the time outside its maximum limits. That means it will draw more current and power out of the electrical power supply than we anticipated in the steady-state design. Once again we would want to make sure the motor could handle this.

You should now modify the program that you used to solve the two simultaneous first-order differential equations and verify my solutions. After that, experiment with the system using your program and study the behavior of the system with a variety of inputs.

An Easy Introduction to Second-Order Systems

What we were really working with in this case study was a second-order system. Two simultaneous (or coupled) first-order differential equations form a second-order linear differential equation with constant coefficients. That's why we haven't found an exact solution to this system—I haven't yet introduced you to second-order systems.

Let's prove that the two simultaneous first-order differential equations we just solved (numerically) really form a second-order differential equation, which has an exact solution. First let's summarize the equations for the motor-table system:

$$m\frac{dv}{dt} + bv = F \qquad (5.11)$$
$$\text{repeated}$$

$$v = r\omega \qquad (5.25)$$
$$\text{repeated}$$

$$J_m\frac{d\omega}{dt} + B\omega = Q_m - Q_L \qquad (5.29)$$
$$\text{repeated}$$

$$Q_L = rF_L \qquad (5.30)$$
$$\text{repeated}$$

$$Q_m = K_1 i_a \qquad (5.39)$$
$$\text{repeated}$$

$$L_a\frac{di_a}{dt} + R_a i_a = E_c - K_2\omega \qquad (5.42)$$
$$\text{repeated}$$

Now begin by substituting (5.25) into (5.11) to eliminate v and then solve for F

$$F = rm\frac{d\omega}{dt} + rb\omega \qquad (5.59)$$

Next, substitute (5.30) and (5.39) into (5.29) to eliminate Q_m and Q_L

$$J_M\frac{d\omega}{dt} + B\omega = K_1 i_a - rF \qquad (5.60)$$

Now substitute (5.59) into (5.60) to eliminate F and collect terms

$$\left(J_M + r^2 m\right)\frac{d\omega}{dt} + \left(B + r^2 b\right)\omega = K_1 i_a \qquad (5.61a)$$

or

$$\tau_{mT}\frac{d\omega}{dt} + \omega = \left(\frac{K_1}{B + r^2 b}\right)i_a \qquad (5.61b)$$

Next divide (5.42) by R_a, substitute τ_a for L_a / R_a, put the result in operational notation and solve for i_a.

$$\frac{L}{R}\frac{di_a}{dt} + i_a = \frac{E_c}{R_a} - \frac{K_2}{R_a}\omega$$

$$\left(\tau_a D + 1\right)i_a = \frac{E_c}{R_a} - \frac{K_2}{R_a}\omega$$

$$i_a = \frac{\dfrac{E_c}{R_a} - \dfrac{K_2}{R_a}\omega}{\tau_a D + 1} \qquad (5.62)$$

Next put (5.61b) into operational notation

$$\left(\tau_{mT}D + 1\right)\omega = \left(\frac{K_1}{B + rb^2}\right)i_a \qquad (5.61c)$$

and substitute (5.62) into (5.61c) to eliminate i_a and then collect terms

$$\left(\tau_{mT}D + 1\right)\omega = \left(\frac{K_1}{B + rb^2}\right)\frac{\dfrac{E_c}{R_a} - \dfrac{K_2}{R_a}\omega}{\left(\tau_a D + 1\right)}$$

$$\left(\tau_{mT}D + 1\right)\left(\tau_a D + 1\right)\omega = \left(\frac{K_1/R_a}{B + r^2 b^2}\right)E_c$$

$$-\left(\frac{K_1 K_2/R_a}{B + r^2 b^2}\right)\omega$$

Multiply the two first-order lags

$$\left[\tau_{mT}\tau_a D^2 + \left(\tau_{mT} + \tau_a\right)D + 1\right]\omega = \left(\frac{K_1/R_a}{B + r^2 b^2}\right)E_c$$

$$-\left(\frac{K_1 K_2/R_a}{B + r^2 b}\right)\omega$$

and place the equation back in standard form

$$\tau_{mT}\tau_a\frac{d^2\omega}{dt^2} + \left(\tau_{mT} + \tau_a\right)\frac{d\omega}{dt}$$

$$+ \left(1 + \frac{K_1 K_2/R_a}{B + r^2 b}\right)\omega = \left(\frac{K_1/R_a}{B + r^2 b}\right)E_c \qquad (5.63)$$

Equation (5.63) is a second-order ordinary linear differential equation with constant coefficients relating the independent (output) variable ω to the independent (input) variable E_c. We can place this result in transfer function form using the operational notation version. That is,

$$\frac{\omega}{E_c} = \frac{\left(\dfrac{K_1/R_a}{B+r^2 b}\right)}{\tau_{mT}\tau_a D^2 + (\tau_{mT}+\tau_a)D + \left(1+\dfrac{K_1 K_2/R_a}{B+r^2 b}\right)}$$

A block diagram of this transfer function is shown in Figure 5.24. Compare this block diagram with the block diagram shown in Figure 5.20—they are equivalent. Both are a complete representation of our motorized table. We'll learn more about second-order differential equations in the next chapter.

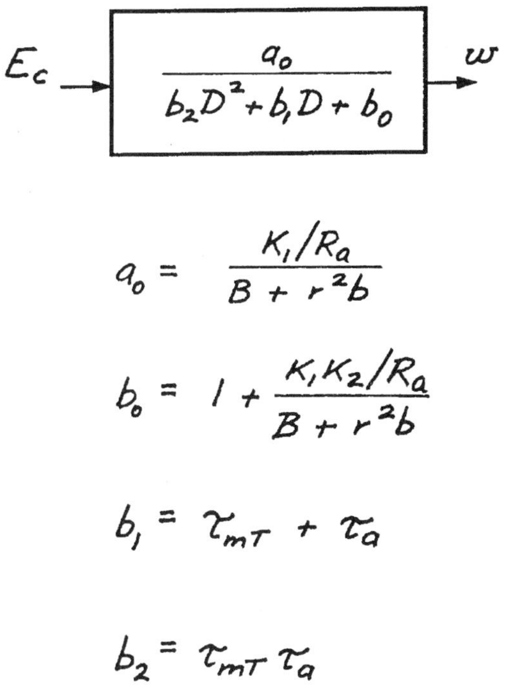

$$a_0 = \frac{K_1/R_a}{B+r^2 b}$$

$$b_0 = 1 + \frac{K_1 K_2/R_a}{B+r^2 b}$$

$$b_1 = \tau_{mT} + \tau_a$$

$$b_2 = \tau_{mT}\tau_a$$

Figure 5.24. Block diagram of transfer function.

Chapter

6 Constructing and Analyzing Second-Order Math Models

Objectives

By the conclusion of this chapter, you will have learned to:

- ■ **Apply the math model tools learned in Chapter 3 to construct models of second-order systems.**

- ■ **Solve second-order math models using numerical techniques.**

- ■ **Analyze second-order math models using exact solution techniques.**

- ■ **Use complex numbers as a "short-cut" for solving math models of second-order systems.**

- ■ **Analyze power in second-order systems.**

6.1 Introduction

In the last three chapters you learned that many real engineering systems can be modeled as a combination of one energy storage and one energy dissipative element. The resulting math model was a first-order linear ordinary differential equation with constant coefficients. You learned that the response of the real system described by such an equation to a given input is governed by a single parameter called the system *time constant*. You found that the time constant was the product or quotient of the coefficients of the energy dissipative and energy storage elements in the system.

In this chapter you will learn how to construct math models of more complex systems. System complexity is dependent on the number of different energy storage elements present in a system. When at least two different energy storage elements are combined, a *second-order* linear ordinary differential equation with constant coefficients is obtained. The response of this equation to a given input is also governed by a single parameter called the *natural frequency* of the system. When a system contains two different energy storage elements and an energy dissipative element, the order of the equation remains the same, but two parameters are required to characterize the output for a given input. These two parameters, called the *natural frequency* and the *damping ratio*, affect the response of the system for a given input.

You will also be introduced to new tools in this chapter which will make it easier to work with more complex math models. The tools can also be used with first-order math models and can eliminate a considerable amount of the algebra we ran into in Chapter 4 when investigating the frequency response of a first-order math model.

6.2 Constructing Second-Order Math Models

The four developmental tools you learned in Chapter 3 are commonly used to develop math models of engineering systems regardless of their complexity. You may find some of these tools easier to use than others when you work with more complex systems. I highly recommend that you use all of the tools, so you can double check your work. You can decide which tool is easier for you to use after you gain experience.

Once again I will first use an electrical system to introduce you to the development and analysis of second-order math models. By now you know it doesn't matter what kind of engineering system I use, since they all look alike mathematically. Figure 6.1 shows an electrical circuit that is similar to the one you worked with so much in Chapters 3 and 4. The only difference between this circuit and the one shown in Figure 3.6 is that an inductor has been added in series with the resistor. The circuit now has two energy storage elements and one energy dissipative element. Let's derive the math model relating the output voltage V_o to the input voltage V_i using all applicable math model development tools.

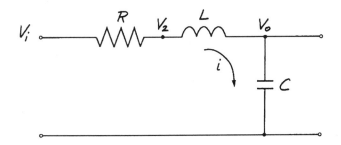

Figure 6.1. Electrical circuit with two energy storage elements and one energy dissipative element.

Path-Vertex-Elemental Equation Method

Begin as we did in Chapter 3 by writing the elemental equation for the resistor

$$i = \frac{1}{R}\left(V_i - V_2\right) \qquad (6.1)$$

Then write the elemental equation for the inductor

$$V_2 - V_o = L\frac{di}{dt} \qquad (6.2)$$

and finally the elemental equation for the capacitor

$$i = C\frac{dV_o}{dt} \qquad (6.3)$$

Now proceed by eliminating unwanted variables. In Chapter 3 the circuit did not have the inductor and we proceeded by substituting (6.3) into (6.1) to eliminate i from the equations. If we do this again we obtain

$$C\frac{dV_o}{dt} = \frac{1}{R}\left(V_i - V_2\right) \qquad (6.4)$$

This gives us a relationship between V_o and V_i, but we also have V_2 involved. We can eliminate V_2 using (6.2), but when we do this, we end up with di/dt in the result and this is not what we wanted.

The procedure to follow when eliminating variables is to use one equation to eliminate the variable *and any of its derivatives* that appear in the other equations. In this example we want to eliminate the current and the derivative of the current from the equations. We do this by using equation (6.3) and the derivative of this equation

$$\frac{di}{dt} = C\frac{d^2V_o}{dt^2}$$

to eliminate i from the other two equations (6.1) and (6.2). We already did part of this and got (6.4). So we complete the elimination of the variable i by substituting the equation for di/dt above into (6.2), giving

$$V_2 - V_o = LC\frac{d^2V_o}{dt^2} \qquad (6.5)$$

So far we have reduced the three equations (6.1), (6.2), and (6.3) to two equations (6.4) and (6.5) by eliminating one of the variables. We do not want the variable V_2 in the final math model, so we reduce the two remaining equations to one by eliminating V_2. Using (6.4) to solve for V_2 gives

$$V_2 = V_i - RC\frac{dV_o}{dt}$$

Substituting this into (6.5) gives

$$V_i - RC\frac{dV_o}{dt} - V_o = LC\frac{d^2V_o}{dt^2}$$

This equation now contains only the output and input variables. It can be rearranged in our standard form where the output variable V_o is on the left side of the equal sign and the input variable V_i is on the right side, resulting in

$$LC\frac{d^2V_o}{dt^2} + RC\frac{dV_o}{dt} + V_o = V_i \qquad (6.6)$$

Equation (6.6) is the desired relationship between V_o and V_i that we were looking for. The math model is called a *second*-order, linear, ordinary differential equation with constant coefficients. It is called that because it contains the output variable and up to the *second derivative* of the output variable.

Note that if L is set to zero, the math model given by (6.6) reduces to the math model given by equation (3.15b). Get used to doing this type of checking. Setting parameters to zero, or sometimes to infinity, is called *limit checking*. It is a good way to see if your math model makes sense. Also note that if R were zero, we would still be left with a *second*-order differential equation even though the first derivative term would be missing.

Remember also to check units after you have derived a math model. If the math model given by equation (6.6) is correct, then every term in the equation must have units of volts. That means LC must have units of time squared and RC must have units of time. We already know RC has units of time. We called RC the *time constant* when only the capacitor and the resis-

tor were in the circuit. Does LC have units of time squared? You know inductors are usually measured in henries and capacitors in farads, but you can see from equation (6.2) that the units of L are

$$[L] = \frac{[V_2 - V_o]}{\left[\dfrac{di}{dt}\right]} = \frac{\text{volts}}{\text{amps / sec}} = \frac{\text{volts} - \text{sec}}{\text{amps}}$$

and from (6.3) you can see the units of C are

$$[C] = \left[\frac{i}{dV_o/dt}\right] = \frac{\text{amps}}{\text{volts / sec}} = \frac{\text{amps} - \text{sec}}{\text{volts}}$$

Consequently, the product LC has units of

$$[L]\times[C] = \frac{\text{volts} - \text{sec}}{\text{amps}} \times \frac{\text{amps} - \text{sec}}{\text{volts}} = \text{sec}^2$$

Impedance Method

Remember when you use this method to follow the steps that were outlined in Chapter 3. First, replace each circuit element with its impedance, as shown in Figure 6.2a.

Treat the impedances as if they were resistors and look for elements to combine that are in series and/or in parallel. In this circuit, Z_R and Z_L are in series and can be combined as shown in Figure 6.2b. Now recognize the simplified circuit as a voltage divider and immediately write the relationship between the output and input voltage as

$$\frac{V_o}{V_i} = \frac{Z_c}{Z_R + Z_L + Z_C} \qquad (6.7)$$

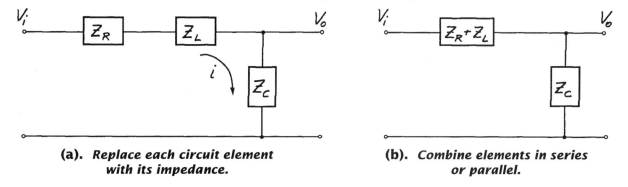

(a). Replace each circuit element with its impedance.

(b). Combine elements in series or parallel.

Figure 6.2. Impedance method.

Next, substitute the elemental impedances

$$\frac{V_o}{V_i} = \frac{1/CD}{R + LD + 1/CD} \qquad (6.8)$$

Simplify to

$$\frac{V_o}{V_i} = \frac{1}{RCD + LCD^2 + 1}$$

Cross-multiply and rearrange terms to get

$$LCD^2 V_o + RCDV_o + V_o = V_i$$

Finally, write the differential equation by substituting $d(\ )/dt$ for operator D, $d^2(\ )/dt^2$ for D^2, etc.

$$LC\frac{d^2 V_o}{dt^2} + RC\frac{dV_o}{dt} + V_o = V_i \qquad (6.9)$$

This result is the same as we obtained using the path-vertex-elemental equation approach, giving us confidence that the math model is correct.

Block Diagram Method

Always start with the elemental equations when using this method and use only integrators in the block diagram. Start the block diagram building process with the output variable on the right side of the diagram produced by, if possible, an integrator block. Since V_o is the desired output, start with equation (6.3), since it has V_o on the right side. It contains the derivative of V_o, and we can rewrite it with an integral as

$$V_o = \frac{1}{C}\int_o^t i\,dt + (V_o)_{init} = \frac{1}{CD}i + (V_o)_{init}$$

Draw this part of the block diagram as shown in Figure 6.3a with V_o as the output and i as the input.

Now we need i as the output of another block to feed into the one we just drew. Once again look for an equation that will produce i as the output of an integrator block, if possible. We can see that equation (6.2) has the derivative of the current and can be rewritten as

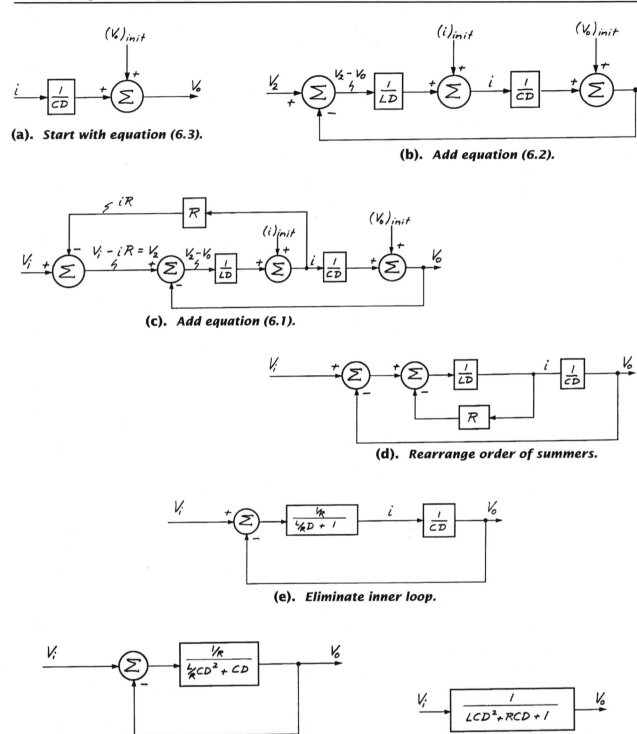

(a). Start with equation (6.3).

(b). Add equation (6.2).

(c). Add equation (6.1).

(d). Rearrange order of summers.

(e). Eliminate inner loop.

(f). Combine blocks.

(g). Transfer function.

Figure 6.3. Block diagrams.

$$i = \frac{1}{L}\int_o^t (V_2 - V_0)\,dt + (i)_{init} = \frac{1}{LD}(V_2 - V_0) + (i)_{init}$$

Draw the block diagram for this integrator next to the previous one using i as the output and V_2 and V_o as the inputs, as shown in Figure 6.3b. Note how V_o was obtained as feedback from the output of the previous integrator block.

Now we need V_2 as the output of a block. We can get this from the final remaining equation (6.1) rewritten as

$$V_2 = V_i - iR$$

This equation can be added to the block diagram as shown in Figure 6.3c. Note how iR is obtained by feeding back i through a block containing R.

Figure 6.3c is a complete symbolic block diagram model of our circuit. We will proceed shortly to develop the math model by reducing this diagram. However, a lot can be learned about the behavior of this circuit just by studying this diagram. Take a moment and compare it with the circuit diagram in Figure 6.1. Assume all the initial conditions are zero and then mentally apply a step change to the input. Sketch each variable against time. I usually like to build a detailed block diagram like this to help me visualize how the engineering system I'm analyzing really works. Equations are great, but they should aide, not supplant, your intuition.

A math model is best developed from a block diagram by reducing the diagram in stages. First, assume the initial conditions for the two integrators are zero. This helps clean up the diagram and makes the reduction easier. Then start

eliminating inner feedback loops. Sometimes these inner feedback loops are not obvious. For example, notice in Figure 6.3c that the order in which the summers operate on variables doesn't matter. Sometimes their order can be interchanged, making inner loops more obvious. The order of the summers in this block diagram can be rearranged as shown in Figure 6.3d. The inner feedback loop now stands out more clearly. This inner loop can now be eliminated, using the block diagram feedback reduction technique explained in Chapter 3. This produces the diagram shown in Figure 6.3e.

Look at Figure 6.3e for a moment and compare it to the circuit diagram shown in Figure 6.1. Notice that the resistor and inductor are in series. Current must therefore pass through both elements before it gets to the capacitor. The resistor and the inductor form a first-order differential equation which we previously dubbed a "lag." Thus, the voltage difference between V_i and V_o produces a current i only after a passage of a certain amount of time. The rate at which current flows into the capacitor is controlled by the time constant L/R and the difference between the input and output voltages.

The two blocks shown in Figure 6.3e can be combined into one as shown in Figure 6.3f. Note that this eliminates current from the block diagram. Using the block diagram feedback reduction technique once again, we arrive at the final block diagram, or transfer function, for the circuit. This transfer function will produce the same differential equation given by the previous two methods. Again we have a confirmation that the math model we derived for this circuit is correct.

Now it would be a good idea to practice developing second-order math models for a variety of engineering systems. Figures 6.4, 6.5, and 6.6 contain electrical, mechanical, and fluid systems whose behavior can be described using

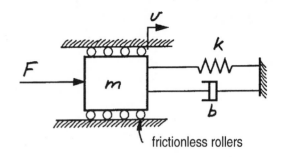

Prepare math model describing the velocity v of the mass as a function of the force F

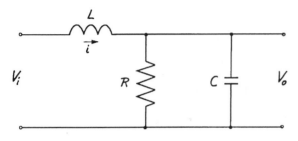

Prepare math model describing V_o as a function of V_i

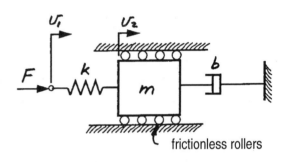

Prepare math model describing the velocity v_2 of the mass as a function of the velocity of the left side of the spring v_1

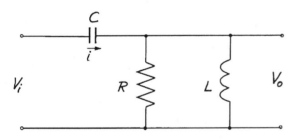

Prepare math model describing V_o as a function of V_i

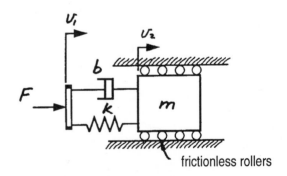

Prepare math model describing the velocity v_2 of the mass as a function of the velocity v_1

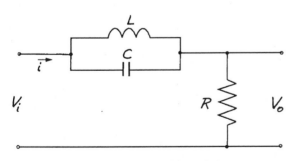

Prepare math model describing V_o as a function of V_i

Figure 6.4. Develop math models for these electrical systems.

Figure 6.5. Develop math models for these mechanical systems.

second-order differential equations. Develop the math model relating the output variable to the input variable as I've indicated in each case. Don't look at the math models I derived in Figures 6.7, 6.8, and 6.9 until you have given your best effort at developing your own. Remember what I said in Chapter 3—the hardest part of engineering system modeling is developing the math models. It takes practice. Once you have the math model, solving it is fairly easy.

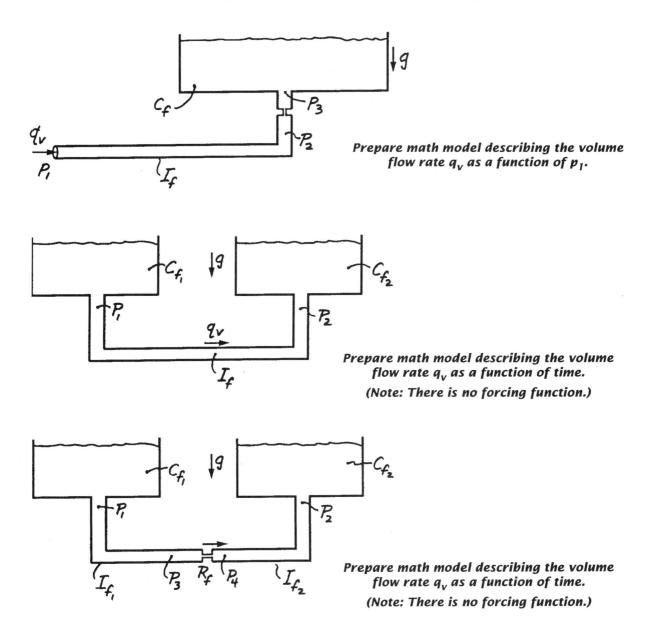

Prepare math model describing the volume flow rate q_v as a function of p_1.

Prepare math model describing the volume flow rate q_v as a function of time.
(Note: There is no forcing function.)

Prepare math model describing the volume flow rate q_v as a function of time.
(Note: There is no forcing function.)

Figure 6.6. Develop math models for these fluid systems.

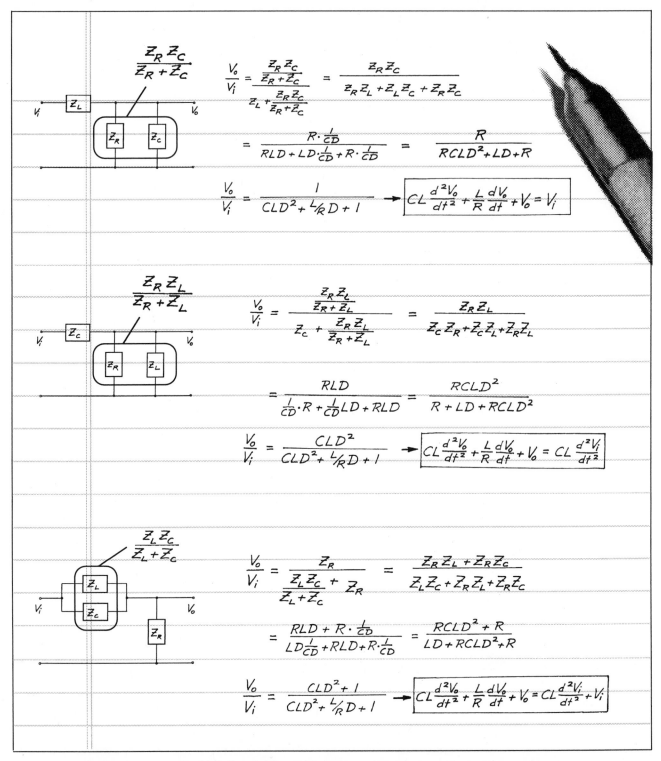

Figure 6.7. Math models for electrical systems shown in Figure 6.4.

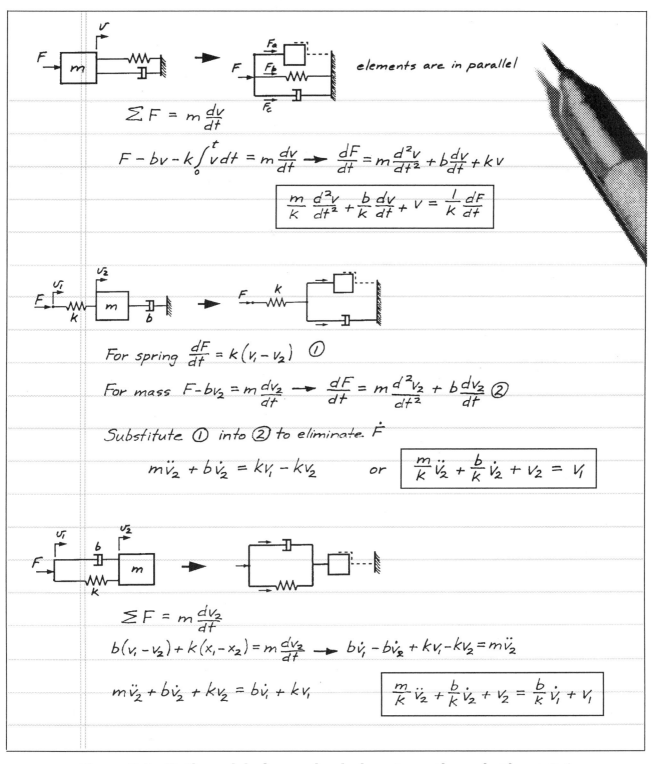

$$\Sigma F = m \frac{dv}{dt}$$

$$F - bv - k\int_0^t v\,dt = m\frac{dv}{dt} \longrightarrow \frac{dF}{dt} = m\frac{d^2v}{dt^2} + b\frac{dv}{dt} + kv$$

$$\boxed{\frac{m}{k}\frac{d^2v}{dt^2} + \frac{b}{k}\frac{dv}{dt} + v = \frac{1}{k}\frac{dF}{dt}}$$

For spring $\dfrac{dF}{dt} = k(v_1 - v_2)$ ①

For mass $F - bv_2 = m\dfrac{dv_2}{dt} \longrightarrow \dfrac{dF}{dt} = m\dfrac{d^2v_2}{dt^2} + b\dfrac{dv_2}{dt}$ ②

Substitute ① into ② to eliminate $\dot{F}$

$$m\ddot{v}_2 + b\dot{v}_2 = kv_1 - kv_2 \qquad \text{or} \qquad \boxed{\frac{m}{k}\ddot{v}_2 + \frac{b}{k}\dot{v}_2 + v_2 = v_1}$$

$$\Sigma F = m\frac{dv_2}{dt}$$

$$b(v_1 - v_2) + k(x_1 - x_2) = m\frac{dv_2}{dt} \longrightarrow b\dot{v}_1 - b\dot{v}_2 + kv_1 - kv_2 = m\ddot{v}_2$$

$$m\ddot{v}_2 + b\dot{v}_2 + kv_2 = b\dot{v}_1 + kv_1 \qquad \boxed{\frac{m}{k}\ddot{v}_2 + \frac{b}{k}\dot{v}_2 + v_2 = \frac{b}{k}\dot{v}_1 + v_1}$$

Figure 6.8. Math models for mechanical systems shown in Figure 6.5.

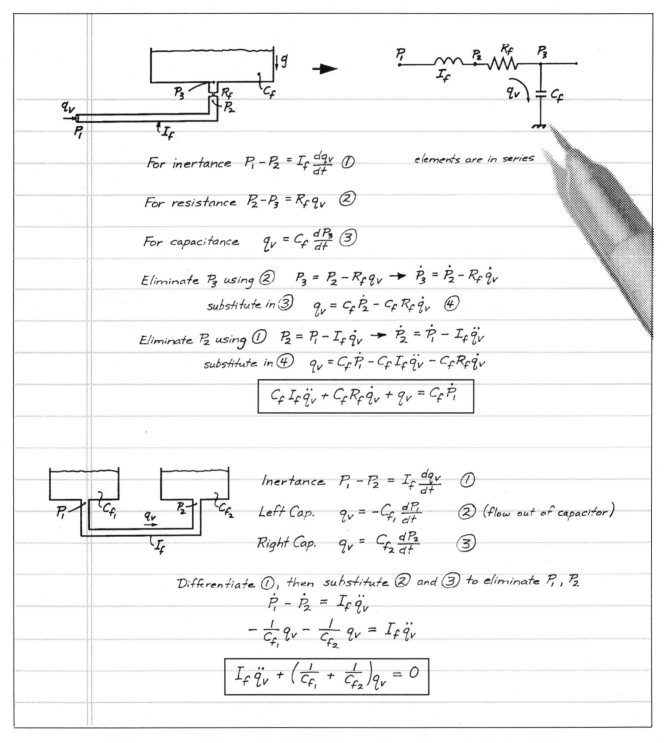

For inertance $\quad P_1 - P_2 = I_f \dfrac{dq_v}{dt}$ ①

For resistance $\quad P_2 - P_3 = R_f q_v$ ②

For capacitance $\quad q_v = C_f \dfrac{dP_3}{dt}$ ③

elements are in series

Eliminate P_3 using ② $\quad P_3 = P_2 - R_f q_v \quad \rightarrow \quad \dot{P}_3 = \dot{P}_2 - R_f \dot{q}_v$

substitute in ③ $\quad q_v = C_f \dot{P}_2 - C_f R_f \dot{q}_v$ ④

Eliminate P_2 using ① $\quad P_2 = P_1 - I_f \dot{q}_v \quad \rightarrow \quad \dot{P}_2 = \dot{P}_1 - I_f \ddot{q}_v$

substitute in ④ $\quad q_v = C_f \dot{P}_1 - C_f I_f \ddot{q}_v - C_f R_f \dot{q}_v$

$$\boxed{C_f I_f \ddot{q}_v + C_f R_f \dot{q}_v + q_v = C_f \dot{P}_1}$$

Inertance $\quad P_1 - P_2 = I_f \dfrac{dq_v}{dt}$ ①

Left Cap. $\quad q_v = -C_{f_1} \dfrac{dP_1}{dt}$ ② (flow out of capacitor)

Right Cap. $\quad q_v = C_{f_2} \dfrac{dP_2}{dt}$ ③

Differentiate ①, then substitute ② and ③ to eliminate P_1, P_2

$$\dot{P}_1 - \dot{P}_2 = I_f \ddot{q}_v$$

$$-\frac{1}{C_{f_1}} q_v - \frac{1}{C_{f_2}} q_v = I_f \ddot{q}_v$$

$$\boxed{I_f \ddot{q}_v + \left(\frac{1}{C_{f_1}} + \frac{1}{C_{f_2}}\right) q_v = 0}$$

Figure 6.9. Math models for fluid systems shown in Figure 6.6.

(continued on next page)

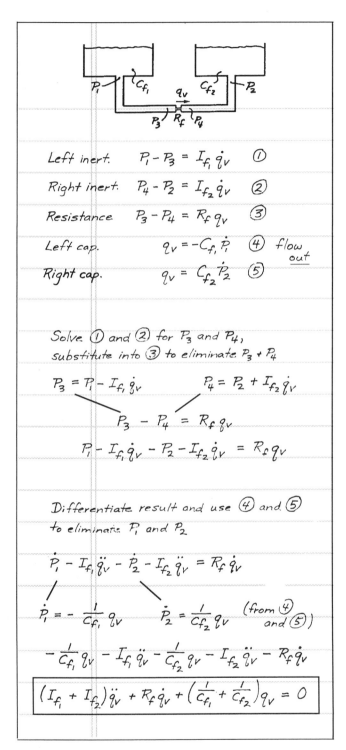

Left inert. $P_1 - P_3 = I_{f_1} \dot{q}_v$ ①

Right inert. $P_4 - P_2 = I_{f_2} \dot{q}_v$ ②

Resistance $P_3 - P_4 = R_f q_v$ ③

Left cap. $q_v = -C_{f_1} \dot{P}_1$ ④ flow out

Right cap. $q_v = C_{f_2} \dot{P}_2$ ⑤

Solve ① and ② for P_3 and P_4, substitute into ③ to eliminate P_3 + P_4

$P_3 = P_1 - I_{f_1} \dot{q}_v$ $P_4 = P_2 + I_{f_2} \dot{q}_v$

$P_3 - P_4 = R_f q_v$

$P_1 - I_{f_1} \dot{q}_v - P_2 - I_{f_2} \dot{q}_v = R_f q_v$

Differentiate result and use ④ and ⑤ to eliminate P_1 and P_2

$\dot{P}_1 - I_{f_1} \ddot{q}_v - \dot{P}_2 - I_{f_2} \ddot{q}_v = R_f \dot{q}_v$

$\dot{P}_1 = -\frac{1}{C_{f_1}} q_v$ $\dot{P}_2 = \frac{1}{C_{f_2}} q_v$ (from ④ and ⑤)

$-\frac{1}{C_{f_1}} q_v - I_{f_1} \ddot{q}_v - \frac{1}{C_{f_2}} q_v - I_{f_2} \ddot{q}_v - R_f \dot{q}_v$

$\boxed{\left(I_{f_1} + I_{f_2}\right) \ddot{q}_v + R_f \dot{q}_v + \left(\frac{1}{C_{f_1}} + \frac{1}{C_{f_2}}\right) q_v = 0}$

Figure 6.9 *(continued)*. Math models for fluid systems shown in Figure 6.6.

6.3 Analyzing Second-Order Math Models Using Numerical Solution Methods

In Chapter 4 you learned how to solve a first-order differential equation numerically. In Chapter 5 you learned how to solve two simultaneous first-order differential equations numerically. You also learned in Chapter 5 that two simultaneous first-order differential equations can be reduced to one second-order differential equation. The reverse is also true. That is, a second-order differential equation can be reduced to two simultaneous first-order differential equations. In fact, an *nth*-order differential equation can be reduced to n simultaneous first-order differential equations.

Since you already know how to solve two simultaneous differential equations numerically, then you *already* know how to solve a second- and higher-order differential equation numerically. You reduce an *nth*-order differential equation to n simultaneous first-order equations by introducing new variables called *dummy* variables. Let me illustrate with our example electrical circuit math model given by equation (6.6). Let

$$u = \frac{dV_o}{dt} \quad \text{or} \quad \frac{dV_o}{dt} = u \qquad (6.10)$$

Now take the derivative of (6.10)

$$\frac{d^2 V_o}{dt^2} = \frac{du}{dt} \qquad (6.11)$$

Substitute (6.10) and (6.11) into (6.6)

$$LC\frac{du}{dt} + RCu + V_o = V_i$$

This is a first-order differential equation with u as the output variable. Rearrange it in the following form:

$$\frac{du}{dt} = \frac{1}{LC}[V_i - V_o - RCu] \qquad (6.12)$$

Now look at equations (6.10) and (6.12). They are two first-order simultaneous differential equations which can be solved numerically using the program you wrote at the end of Chapter 5.

If the use of dummy variables seems too hokey or mathematically abstract, look at it this way: The second term in equation (6.6) is equal to Ri. That is,

$$RC\frac{dV_o}{dt} = Ri$$

because

$$C\frac{dV_o}{dt} = i$$

Consequently, we can write this elemental equation in the form of a first-order differential equation

$$\frac{dV_o}{dt} = \frac{1}{C}i \qquad (6.13)$$

We can now express the second derivative of the output voltage as

$$\frac{d^2V_o}{dt^2} = \frac{1}{C}\frac{di}{dt} \qquad (6.14)$$

Then substitute (6.13) and (6.14) into (6.6)

$$L\frac{di}{dt} + Ri + V_o = V_i$$

and rearrange it in the form

$$\frac{di}{dt} = \frac{1}{L}[V_i - V_o - Ri] \qquad (6.15)$$

Equations (6.13) and (6.15) are now two simultaneous first-order differential equations similar to those given by equations (6.10) and (6.12).

Note that equations (6.13) and (6.15) correspond to the two integrator blocks shown in Figure 6.3(c). This is not coincidental. By arranging the block diagram so that only integrators are used, the first-order simultaneous differential equations can be written directly from the block diagram. Clearly the number of differentiators in the block diagram equals the order of the describing differential equation.

Following the procedures described above, any *nth*-order differential equation can be written in the form of n simultaneous first-order differential equations. You simply follow the procedure described by equations (6.10) and (6.11) and keep introducing dummy variables until only a first-order differential equation remains when all of the dummy variables are substituted for the derivatives in the original *nth*-order differential equation. The procedure works for nonlinear, as well as linear, ordinary differential equations.

When you learned to numerically solve the two simultaneous first-order differential equations in Chapter 4, you used for the time step 1/10th of the smaller time constant. Now what do you use? My rule is to use about 1/20th of the coefficient that goes with the first derivative term when the equation is rearranged so that the coefficient of the output variable is one. So for equation (6.6), which is already in this form, I would use a time step equal to $RC/20$.

Response to a Step Input

Now solve the math model given by equation (6.6) numerically using a step input for V_i and using the program you developed at the end of Chapter 5. Use the following:

$$t_{end} = 15 \text{ seconds}$$
$$L = 1 \text{ henry}$$
$$C = 1 \text{ farad}$$
$$R = 0.6 \text{ ohm}$$
$$(i)_{init} = 0 \text{ [equivalent to } (dV_o / dt)_{init} = 0]$$
$$(V_o)_{init} = 0$$
$$V_{is} = 10 \text{ volts}$$

After you complete your work, compare your results with mine shown in Table 6.1 and Figure 6.10. I obtained my solution using an Excel spreadsheet. I solved the equations in the form given by (6.13) and (6.15) so I could plot i and V_o easily.

The response of this circuit to a step input is quite interesting. The output voltage first rises slowly and then very quickly. It rises so quickly in fact that it overshoots the input voltage. Then it reverses direction and undershoots. It continues to do this for several cycles. Each time the overshoot and undershoot is less and less.

Now repeat the solution with R increased to 1.0 and then to 1.4 ohms. Compare your results with mine given in Tables 6.2 and 6.3 and Figures 6.11 and 6.12, respectively. You can see that the overshoot and undershoot phenomena decreases as the value of the resistor is increased. What is going on here is that energy is being alternately stored in the inductor and then in the capacitor during the overshoot-undershoot cycles. At time t = 0, the input voltage goes from 0 to 10 volts. Energy in excess of that dissi-

pated by the resistor in steady state is pumped into the system and stored in the inductor and then in the capacitor. As the output voltage builds, energy from the inductor is used to maintain the current in the inductor, which in turn overcharges the capacitor. The output voltage therefore exceeds the input voltage and this causes the current in the inductor to reverse. Then excess energy is taken from the capacitor and the output voltage drops below the input voltage, causing the current in the inductor to reverse again. The excess energy initially pumped into the system is eventually dissipated by the resistor. The larger the resistor, the faster this excess energy is dissipated.

If the value of the resistor is made smaller and smaller, the magnitude of the oscillations will increase. When the value of the resistor equals zero, the system oscillates continuously at a frequency determined by the values of L and C. Set the value of R equal to zero in your program and repeat your solution. Compare your results with mine given in Table 6.4 and Figure 6.13. Clearly when R is set to zero, the oscillations do not decay. Without any way of dissipating the excess energy pumped initially into the system, the oscillations will go on forever. Remember that we are talking about ideal elements. That means ideal inductors and capacitors have no resistance.

Now let's investigate the effect of L and C on the frequency of the oscillations. Set L and C to 0.5 and then to 2.0 and repeat the solutions. Compare your results with mine given in Tables 6.5 and 6.6 and Figures 6.14 and 6.15. Clearly, increasing L and C has decreased the frequency of oscillation, whereas decreasing L and C has increased the frequency.

Table 6.1.

L	1					
C	1					
R	0.6					
(V)init	0					
(i)init	0					
Vin*	10					
Tend	3					
Δt	0.025					
t	i	V	di/dt	Δi	dV/dt	ΔV
0	=B5	=B4	=($B$6-C11-$B$3*B11)/$B$1	=D11*$B$8	=B11/$B$2	=F11*$B$8
=A11+$B$8	=B11+E11	=C11+G11	=($B$6-C12-$B$3*B12)/$B$1	=D12*$B$8	=B12/$B$2	=F12*$B$8
=A12+$B$8	=B12+E12	=C12+G12	=($B$6-C13-$B$3*B13)/$B$1	=D13*$B$8	=B13/$B$2	=F13*$B$8
=A13+$B$8	=B13+E13	=C13+G13	=($B$6-C14-$B$3*B14)/$B$1	=D14*$B$8	=B14/$B$2	=F14*$B$8
=A14+$B$8	=B14+E14	=C14+G14	=($B$6-C15-$B$3*B15)/$B$1	=D15*$B$8	=B15/$B$2	=F15*$B$8
=A15+$B$8	=B15+E15	=C15+G15	=($B$6-C16-$B$3*B16)/$B$1	=D16*$B$8	=B16/$B$2	=F16*$B$8
=A16+$B$8	=B16+E16	=C16+G16	=($B$6-C17-$B$3*B17)/$B$1	=D17*$B$8	=B17/$B$2	=F17*$B$8
=A17+$B$8	=B17+E17	=C17+G17	=($B$6-C18-$B$3*B18)/$B$1	=D18*$B$8	=B18/$B$2	=F18*$B$8

L	1					
C	1					
R	0.6					
(V)init	0					
(i)init	0					
Vin*	10					
Tend	3					
Δt	0.025					
t	i	V	di/dt	Δi	dV/dt	ΔV
0	0	0	10	0.2500	0	0.0000
0.03	0.25	0.00	9.85	0.2463	0.25	0.0063
0.05	0.50	0.01	9.70	0.2424	0.50	0.0124
0.08	0.74	0.02	9.54	0.2385	0.74	0.0185
0.10	0.98	0.04	9.38	0.2344	0.98	0.0244
0.13	1.21	0.06	9.21	0.2303	1.21	0.0303
0.15	1.44	0.09	9.04	0.2261	1.44	0.0360
0.18	1.67	0.13	8.87	0.2218	1.67	0.0417
0.20	1.89	0.17	8.70	0.2174	1.89	0.0472
0.23	2.11	0.22	8.52	0.2130	2.11	0.0527
0.25	2.32	0.27	8.34	0.2085	2.32	0.0580
0.28	2.53	0.33	8.16	0.2039	2.53	0.0632
0.30	2.73	0.39	7.97	0.1992	2.73	0.0683
0.33	2.93	0.46	7.78	0.1945	2.93	0.0733
0.35	3.13	0.53	7.59	0.1898	3.13	0.0782
0.38	3.32	0.61	7.40	0.1850	3.32	0.0829
0.40	3.50	0.69	7.21	0.1802	3.50	0.0875

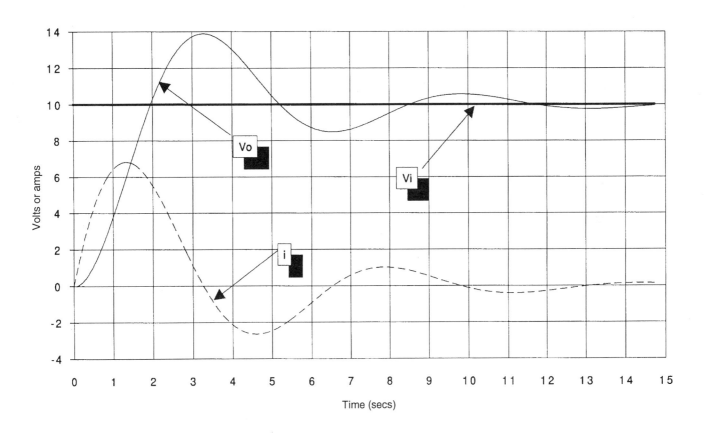

Figure 6.10.
Step response of circuit in Figure 6.1 with R = 0.6 ohm, L = C = 1.

Table 6.2.

L	1					
C	1					
R	1					
(V)init	0					
(i)init	0					
Vin*	10					
Tend	3					
Δt	0.025					
t	i	V	di/dt	Δi	dV/dt	ΔV
0	0	0	10	0.2500	0	0.0000
0.03	0.25	0.00	9.75	0.2438	0.25	0.0063
0.05	0.49	0.01	9.50	0.2375	0.49	0.0123
0.08	0.73	0.02	9.25	0.2313	0.73	0.0183
0.10	0.96	0.04	9.00	0.2250	0.96	0.0241
0.13	1.19	0.06	8.75	0.2188	1.19	0.0297
0.15	1.41	0.09	8.50	0.2126	1.41	0.0352
0.18	1.62	0.13	8.26	0.2064	1.62	0.0405
0.20	1.83	0.17	8.01	0.2002	1.83	0.0456
0.23	2.03	0.21	7.76	0.1941	2.03	0.0506
0.25	2.22	0.26	7.52	0.1879	2.22	0.0555
0.28	2.41	0.32	7.27	0.1819	2.41	0.0602
0.30	2.59	0.38	7.03	0.1758	2.59	0.0647
0.33	2.77	0.44	6.79	0.1698	2.77	0.0691
0.35	2.93	0.51	6.55	0.1638	2.93	0.0734
0.38	3.10	0.59	6.32	0.1579	3.10	0.0775
0.40	3.26	0.66	6.08	0.1520	3.26	0.0814
0.43	3.41	0.74	5.85	0.1462	3.41	0.0852
0.45	3.55	0.83	5.62	0.1404	3.55	0.0889
0.48	3.70	0.92	5.39	0.1347	3.70	0.0924
0.50	3.83	1.01	5.16	0.1290	3.83	0.0957
0.53	3.96	1.11	4.93	0.1234	3.96	0.0990
0.55	4.08	1.21	4.71	0.1178	4.08	0.1021
0.58	4.20	1.31	4.49	0.1123	4.20	0.1050
0.60	4.31	1.41	4.28	0.1069	4.31	0.1078
0.63	4.42	1.52	4.06	0.1015	4.42	0.1105
0.65	4.52	1.63	3.85	0.0962	4.52	0.1130
0.68	4.62	1.74	3.64	0.0910	4.62	0.1154
0.70	4.71	1.86	3.43	0.0858	4.71	0.1177
0.73	4.79	1.98	3.23	0.0807	4.79	0.1198
0.75	4.87	2.10	3.03	0.0757	4.87	0.1219
0.78	4.95	2.22	2.83	0.0708	4.95	0.1238
0.80	5.02	2.34	2.64	0.0659	5.02	0.1255
0.83	5.09	2.47	2.45	0.0611	5.09	0.1272
0.85	5.15	2.60	2.26	0.0564	5.15	0.1287
0.88	5.20	2.72	2.07	0.0518	5.20	0.1301
0.90	5.26	2.85	1.89	0.0472	5.26	0.1314
0.93	5.30	2.99	1.71	0.0428	5.30	0.1326
0.95	5.35	3.12	1.54	0.0384	5.35	0.1337

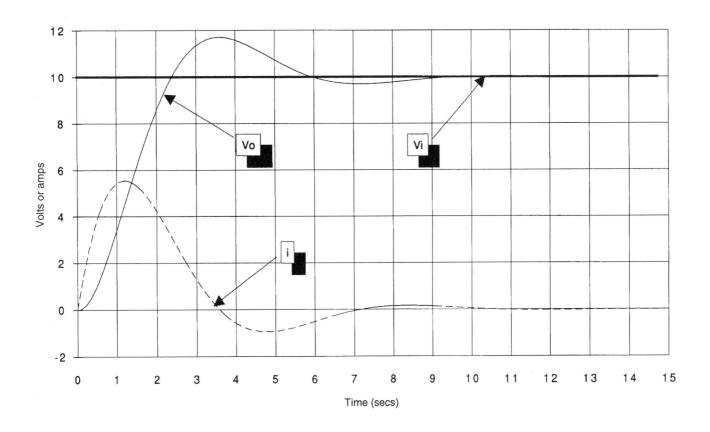

Figure 6.11.
Step response of circuit in Figure 6.1 with R = 1.0 ohm, L = C = 1.

Table 6.3.

L	1					
C	1					
R	1.4					
(V)init	0					
(i)init	0					
Vin*	10					
Tend	3					
Δt	0.025					
t	i	V	di/dt	Δi	dV/dt	ΔV
0	0	0	10	0.2500	0	0.0000
0.03	0.25	0.00	9.65	0.2413	0.25	0.0063
0.05	0.49	0.01	9.31	0.2327	0.49	0.0123
0.08	0.72	0.02	8.97	0.2242	0.72	0.0181
0.10	0.95	0.04	8.64	0.2159	0.95	0.0237
0.13	1.16	0.06	8.31	0.2078	1.16	0.0291
0.15	1.37	0.09	7.99	0.1998	1.37	0.0343
0.18	1.57	0.12	7.68	0.1919	1.57	0.0393
0.20	1.76	0.16	7.37	0.1842	1.76	0.0441
0.23	1.95	0.21	7.07	0.1767	1.95	0.0487
0.25	2.12	0.26	6.77	0.1693	2.12	0.0531
0.28	2.29	0.31	6.48	0.1620	2.29	0.0573
0.30	2.46	0.37	6.20	0.1549	2.46	0.0614
0.33	2.61	0.43	5.92	0.1479	2.61	0.0653
0.35	2.76	0.49	5.65	0.1411	2.76	0.0690
0.38	2.90	0.56	5.38	0.1345	2.90	0.0725
0.40	3.03	0.63	5.12	0.1280	3.03	0.0758
0.43	3.16	0.71	4.86	0.1216	3.16	0.0790
0.45	3.28	0.79	4.61	0.1153	3.28	0.0821
0.48	3.40	0.87	4.37	0.1093	3.40	0.0850
0.50	3.51	0.96	4.13	0.1033	3.51	0.0877
0.53	3.61	1.04	3.90	0.0975	3.61	0.0903
0.55	3.71	1.13	3.67	0.0918	3.71	0.0927
0.58	3.80	1.23	3.45	0.0863	3.80	0.0950
0.60	3.89	1.32	3.24	0.0809	3.89	0.0972
0.63	3.97	1.42	3.03	0.0756	3.97	0.0992
0.65	4.04	1.52	2.82	0.0705	4.04	0.1011
0.68	4.11	1.62	2.62	0.0655	4.11	0.1029
0.70	4.18	1.72	2.43	0.0607	4.18	0.1045
0.73	4.24	1.83	2.24	0.0559	4.24	0.1060
0.75	4.30	1.93	2.05	0.0513	4.30	0.1074
0.78	4.35	2.04	1.87	0.0468	4.35	0.1087
0.80	4.39	2.15	1.70	0.0425	4.39	0.1099
0.83	4.44	2.26	1.53	0.0382	4.44	0.1109
0.85	4.48	2.37	1.37	0.0341	4.48	0.1119
0.88	4.51	2.48	1.21	0.0301	4.51	0.1127
0.90	4.54	2.59	1.05	0.0263	4.54	0.1135
0.93	4.57	2.71	0.90	0.0225	4.57	0.1141
0.95	4.59	2.82	0.75	0.0189	4.59	0.1147

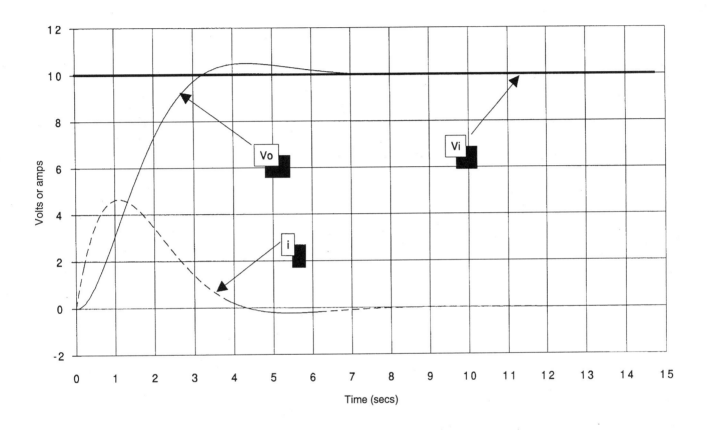

Figure 6.12.
Step response of circuit in Figure 6.1 with R = 1.4 ohm, L = C = 1.

Table 6.4.

L	1					
C	1					
R	0					
(V)init	0					
(i)init	0					
Vin*	10					
Tend	3					
Δt	0.025					
t	i	V	di/dt	Δi	dV/dt	ΔV
0	0	0	10	0.2500	0	0.0000
0.03	0.25	0.00	10.00	0.2500	0.25	0.0063
0.05	0.50	0.01	9.99	0.2498	0.50	0.0125
0.08	0.75	0.02	9.98	0.2495	0.75	0.0187
0.10	1.00	0.04	9.96	0.2491	1.00	0.0250
0.13	1.25	0.06	9.94	0.2484	1.25	0.0312
0.15	1.50	0.09	9.91	0.2477	1.50	0.0374
0.18	1.74	0.13	9.87	0.2467	1.74	0.0436
0.20	1.99	0.17	9.83	0.2456	1.99	0.0498
0.23	2.24	0.22	9.78	0.2444	2.24	0.0559
0.25	2.48	0.28	9.72	0.2430	2.48	0.0620
0.28	2.72	0.34	9.66	0.2414	2.72	0.0681
0.30	2.97	0.41	9.59	0.2397	2.97	0.0741
0.33	3.21	0.48	9.52	0.2379	3.21	0.0801
0.35	3.44	0.56	9.44	0.2359	3.44	0.0861
0.38	3.68	0.65	9.35	0.2337	3.68	0.0920
0.40	3.91	0.74	9.26	0.2314	3.91	0.0978
0.43	4.14	0.84	9.16	0.2290	4.14	0.1036
0.45	4.37	0.94	9.06	0.2264	4.37	0.1093
0.48	4.60	1.05	8.95	0.2237	4.60	0.1150
0.50	4.82	1.17	8.83	0.2208	4.82	0.1206
0.53	5.04	1.29	8.71	0.2178	5.04	0.1261
0.55	5.26	1.42	8.58	0.2146	5.26	0.1315
0.58	5.48	1.55	8.45	0.2113	5.48	0.1369
0.60	5.69	1.68	8.32	0.2079	5.69	0.1422
0.63	5.90	1.83	8.17	0.2043	5.90	0.1474
0.65	6.10	1.97	8.03	0.2007	6.10	0.1525
0.68	6.30	2.13	7.87	0.1969	6.30	0.1575
0.70	6.50	2.28	7.72	0.1929	6.50	0.1624
0.73	6.69	2.45	7.55	0.1889	6.69	0.1673
0.75	6.88	2.61	7.39	0.1847	6.88	0.1720
0.78	7.06	2.79	7.21	0.1804	7.06	0.1766
0.80	7.24	2.96	7.04	0.1760	7.24	0.1811
0.83	7.42	3.14	6.86	0.1714	7.42	0.1855
0.85	7.59	3.33	6.67	0.1668	7.59	0.1898
0.88	7.76	3.52	6.48	0.1620	7.76	0.1940
0.90	7.92	3.71	6.29	0.1572	7.92	0.1980
0.93	8.08	3.91	6.09	0.1522	8.08	0.2019
0.95	8.23	4.11	5.89	0.1472	8.23	0.2058

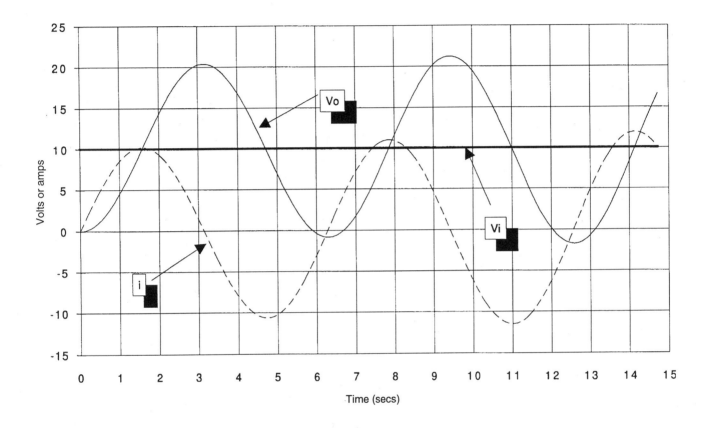

Figure 6.13.
Step response of circuit in Figure 6.1 with R = 0 ohm, L = C = 1.

Table 6.5.

L	0.5					
C	0.5					
R	0					
(V)init	0					
(i)init	0					
Vin*	10					
Tend	3					
Δt	0.025					
t	i	V	di/dt	Δi	dV/dt	ΔV
0	0	0	20	0.5000	0	0.0000
0.03	0.50	0.00	20.00	0.5000	1.00	0.0250
0.05	1.00	0.03	19.95	0.4988	2.00	0.0500
0.08	1.50	0.08	19.85	0.4963	3.00	0.0749
0.10	2.00	0.15	19.70	0.4925	3.99	0.0998
0.13	2.49	0.25	19.50	0.4875	4.98	0.1244
0.15	2.98	0.37	19.25	0.4813	5.95	0.1488
0.18	3.46	0.52	18.95	0.4739	6.91	0.1728
0.20	3.93	0.70	18.61	0.4652	7.86	0.1965
0.23	4.40	0.89	18.22	0.4554	8.79	0.2198
0.25	4.85	1.11	17.78	0.4444	9.70	0.2425
0.28	5.30	1.35	17.29	0.4323	10.59	0.2648
0.30	5.73	1.62	16.76	0.4190	11.45	0.2864
0.33	6.15	1.91	16.19	0.4047	12.29	0.3073
0.35	6.55	2.21	15.57	0.3894	13.10	0.3276
0.38	6.94	2.54	14.92	0.3730	13.88	0.3470
0.40	7.31	2.89	14.23	0.3556	14.63	0.3657
0.43	7.67	3.25	13.49	0.3373	15.34	0.3835
0.45	8.01	3.64	12.73	0.3182	16.01	0.4003
0.48	8.32	4.04	11.93	0.2982	16.65	0.4162
0.50	8.62	4.45	11.09	0.2773	17.25	0.4311
0.53	8.90	4.88	10.23	0.2558	17.80	0.4450
0.55	9.16	5.33	9.34	0.2335	18.31	0.4578
0.58	9.39	5.79	8.43	0.2106	18.78	0.4695
0.60	9.60	6.26	7.49	0.1872	19.20	0.4800
0.63	9.79	6.74	6.53	0.1632	19.57	0.4894
0.65	9.95	7.23	5.55	0.1387	19.90	0.4975
0.68	10.09	7.72	4.55	0.1138	20.18	0.5045
0.70	10.20	8.23	3.54	0.0886	20.41	0.5102
0.73	10.29	8.74	2.52	0.0631	20.58	0.5146
0.75	10.35	9.25	1.49	0.0374	20.71	0.5177
0.78	10.39	9.77	0.46	0.0115	20.78	0.5196
0.80	10.40	10.29	-0.58	-0.0145	20.81	0.5202
0.83	10.39	10.81	-1.62	-0.0405	20.78	0.5195
0.85	10.35	11.33	-2.66	-0.0665	20.70	0.5174
0.88	10.28	11.85	-3.69	-0.0924	20.56	0.5141
0.90	10.19	12.36	-4.72	-0.1181	20.38	0.5095
0.93	10.07	12.87	-5.74	-0.1435	20.14	0.5036
0.95	9.93	13.37	-6.75	-0.1687	19.86	0.4964

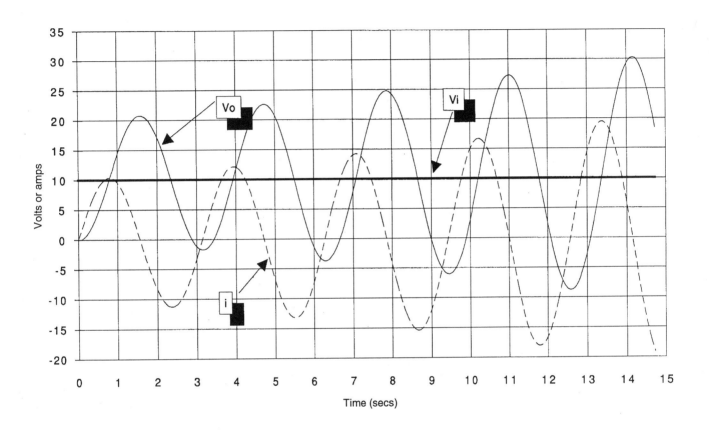

Figure 6.14.
Step response of circuit in Figure 6.1 with R = 0 ohm, L = C = 0.5.

Table 6.6.

L	2					
C	2					
R	0					
(V)init	0					
(i)init	0					
Vin*	10					
Tend	3					
Δt	0.025					
t	i	V	di/dt	Δi	dV/dt	ΔV
0	0	0	5	0.1250	0	0.0000
0.03	0.13	0.00	5.00	0.1250	0.06	0.0016
0.05	0.25	0.00	5.00	0.1250	0.13	0.0031
0.08	0.37	0.00	5.00	0.1249	0.19	0.0047
0.10	0.50	0.01	5.00	0.1249	0.25	0.0062
0.13	0.62	0.02	4.99	0.1248	0.31	0.0078
0.15	0.75	0.02	4.99	0.1247	0.37	0.0094
0.18	0.87	0.03	4.98	0.1246	0.44	0.0109
0.20	1.00	0.04	4.98	0.1245	0.50	0.0125
0.23	1.12	0.06	4.97	0.1243	0.56	0.0140
0.25	1.25	0.07	4.96	0.1241	0.62	0.0156
0.28	1.37	0.09	4.96	0.1239	0.69	0.0171
0.30	1.50	0.10	4.95	0.1237	0.75	0.0187
0.33	1.62	0.12	4.94	0.1235	0.81	0.0202
0.35	1.74	0.14	4.93	0.1232	0.87	0.0218
0.38	1.87	0.16	4.92	0.1230	0.93	0.0233
0.40	1.99	0.19	4.91	0.1227	0.99	0.0249
0.43	2.11	0.21	4.89	0.1224	1.06	0.0264
0.45	2.23	0.24	4.88	0.1220	1.12	0.0279
0.48	2.36	0.27	4.87	0.1217	1.18	0.0295
0.50	2.48	0.30	4.85	0.1213	1.24	0.0310
0.53	2.60	0.33	4.84	0.1209	1.30	0.0325
0.55	2.72	0.36	4.82	0.1205	1.36	0.0340
0.58	2.84	0.39	4.80	0.1201	1.42	0.0355
0.60	2.96	0.43	4.79	0.1196	1.48	0.0370
0.63	3.08	0.47	4.77	0.1192	1.54	0.0385
0.65	3.20	0.50	4.75	0.1187	1.60	0.0400
0.68	3.32	0.54	4.73	0.1182	1.66	0.0415
0.70	3.44	0.59	4.71	0.1177	1.72	0.0430
0.73	3.55	0.63	4.69	0.1171	1.78	0.0444
0.75	3.67	0.67	4.66	0.1166	1.84	0.0459
0.78	3.79	0.72	4.64	0.1160	1.89	0.0473
0.80	3.90	0.77	4.62	0.1154	1.95	0.0488
0.83	4.02	0.82	4.59	0.1148	2.01	0.0502
0.85	4.13	0.87	4.57	0.1142	2.07	0.0517
0.88	4.25	0.92	4.54	0.1135	2.12	0.0531
0.90	4.36	0.97	4.51	0.1129	2.18	0.0545
0.93	4.47	1.02	4.49	0.1122	2.24	0.0559
0.95	4.59	1.08	4.46	0.1115	2.29	0.0573

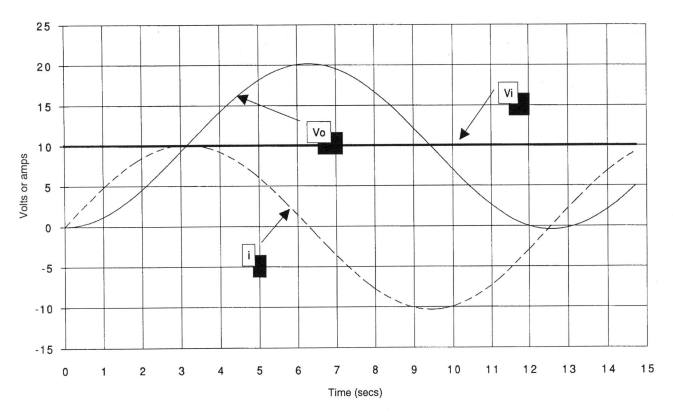

Figure 6.15.
Step response of circuit in Figure 6.1 with R = 0 ohm, L = C = 2.

There is a relationship between the frequency of the oscillations and *L* and *C*. You may want to experiment with other values of *L* and *C* to see if you can determine the relationship. The frequency of oscillation of any second-order math model without an energy dissipative device is called the *natural frequency* of the system. Such a system is easily excited by the least disturbance and will oscillate "forever" at its natural frequency. A weight on a spring is a system that can be described by a second-order differential equation. Tie a pencil to a rubber band and watch the motion of the pencil as you let it fall. The pencil loses very little energy due to air friction and once set in motion it will oscillate for a long time at a frequency determined by the weight of the pencil and the stiffness of the rubber band. Radios used to make use of an inductor and capacitor in parallel. The natural frequency of this circuit was used to send and receive information at a radio frequency equal to the natural frequency. We will be investigating ways to determine the natural frequency shortly.

Response to a Sinusoidal Input

You will recall from Chapter 4 that we investigated the *frequency response* of a first-order system by subjecting it to an input function of the form

$$V_i = V_{is} \sin(2\pi ft)$$

where V_{is} is the amplitude of a sinusoidal input and f is the frequency. You will also recall that after the initial transient died out, the output had the same frequency as the input, the amplitude of the output was less than the amplitude of the input, and the output lagged behind the input. We plotted the ratio of the output and input amplitudes and their phase shift as a function of the input frequency, and we called this the *frequency response* of the system.

Now solve the math model given by equation (6.6) numerically using a sinusoidal input for V_i. Use the following:

$$t_{end} = 15 \text{ seconds}$$
$$L = 1 \text{ henry}$$
$$C = 1 \text{ farad}$$
$$R = 0.6 \text{ ohm}$$
$$(i)_{init} = 0 \text{ [equivalent to } (dV_o / dt)_{init} = 0]$$
$$(V_o)_{init} = 0$$
$$V_{is} = 10 \text{ volts}$$
$$f = 0.05, 0.10, 0.15, 0.20 \text{ and } 0.25 \text{ Hz.}$$

After you obtain the results, calculate the amplitude ratio and the phase shift and plot this against the frequency. Compute the amplitude ratio and phase shift after around 4 or 5 cycles of the input sinusoid have elapsed.

Now compare your results with mine given in Table 6.7 and Figure 6.16. Note that as the frequency increases, the amplitude ratio does not continually decrease as it did with a first-order system. There is a point approximately around 0.15 Hz where the amplitude of the output signal is actually greater than the amplitude of the input signal. As the frequency of the input signal increases past this point, the amplitude ratio starts decreasing and looking more like that for the first-order system. Also note the phase shift. With a first-order system we saw that the phase shift never lagged by more than 90 degrees. With this second-order system, it appears that the phase lag is approaching 180 degrees. You may want to experiment with your model some more to further investigate the output amplitude amplification phenomena. Repeat the solutions with R equal to 0.3 ohms and $R = 1.0$ ohms.

6.4 Analyzing Second-Order Math Models Using Exact Solution Methods

In Chapter 4 you learned that the complete exact solution of a first-order, linear, ordinary differential equation is made up of two parts called the *homogeneous* and *particular* solutions. When the differential equation is placed in its proper form, as is equation (6.6), then the homogeneous solution is obtained by solving the differential equation with the left side (called the *characteristic equation*) set to zero. The particular solution, you will recall, can be determined by the method of undetermined coefficients. In this method a function that looks similar to the input (forcing) function is chosen and then substituted into the differential equation using unknown coefficients. The unknown

Table 6.7

L	1					
C	1					
R	0.6					
(V)init	0					
(i)init	0					
Vin*	10	*Sin (2πft)				
Tend	3					
Δt	0.025					
f	0.25					
t	i	V	di/dt		Δi	dV/dt	ΔV
0	=B5	=B4	=($B$6*SIN(2*PI()*$B$9*A11)-C11-$B$3*B11)/$B$1	=D11*$B$8	=B11/$B$2	=F11*$B$8
=A11+$B$8	=B11+E11	=C11+G11	=($B$6*SIN(2*PI()*$B$9*A12)-C12-$B$3*B12)/$B$1	=D12*$B$8	=B12/$B$2	=F12*$B$8
=A12+$B$8	=B12+E12	=C12+G12	=($B$6*SIN(2*PI()*$B$9*A13)-C13-$B$3*B13)/$B$1	=D13*$B$8	=B13/$B$2	=F13*$B$8
=A13+$B$8	=B13+E13	=C13+G13	=($B$6*SIN(2*PI()*$B$9*A14)-C14-$B$3*B14)/$B$1	=D14*$B$8	=B14/$B$2	=F14*$B$8
=A14+$B$8	=B14+E14	=C14+G14	=($B$6*SIN(2*PI()*$B$9*A15)-C15-$B$3*B15)/$B$1	=D15*$B$8	=B15/$B$2	=F15*$B$8
=A15+$B$8	=B15+E15	=C15+G15	=($B$6*SIN(2*PI()*$B$9*A16)-C16-$B$3*B16)/$B$1	=D16*$B$8	=B16/$B$2	=F16*$B$8
=A16+$B$8	=B16+E16	=C16+G16	=($B$6*SIN(2*PI()*$B$9*A17)-C17-$B$3*B17)/$B$1	=D17*$B$8	=B17/$B$2	=F17*$B$8
=A17+$B$8	=B17+E17	=C17+G17	=($B$6*SIN(2*PI()*$B$9*A18)-C18-$B$3*B18)/$B$1	=D18*$B$8	=B18/$B$2	=F18*$B$8

L	1					
C	1					
R	0.6					
(V)init	0					
(i)init	0					
Vin*	10	*Sin (2πft)				
Tend	3					
Δt	0.025					
f	0.25					
t	i	V	di/dt	Δi	dV/dt	ΔV
0	0	0	0.00	0	0	0
0.03	0.00	0.00	0.39	0.01	0.00	0.00
0.05	0.01	0.00	0.78	0.02	0.01	0.00
0.08	0.03	0.00	1.16	0.03	0.03	0.00
0.10	0.06	0.00	1.53	0.04	0.06	0.00
0.13	0.10	0.00	1.89	0.05	0.10	0.00
0.15	0.14	0.00	2.24	0.06	0.14	0.00
0.18	0.20	0.01	2.59	0.06	0.20	0.00
0.20	0.26	0.01	2.92	0.07	0.26	0.01
0.23	0.34	0.02	3.24	0.08	0.34	0.01
0.25	0.42	0.03	3.55	0.09	0.42	0.01
0.28	0.51	0.04	3.84	0.10	0.51	0.01
0.30	0.60	0.05	4.13	0.10	0.60	0.02
0.33	0.71	0.07	4.40	0.11	0.71	0.02
0.35	0.82	0.08	4.65	0.12	0.82	0.02
0.38	0.93	0.10	4.89	0.12	0.93	0.02
0.40	1.05	0.13	5.12	0.13	1.05	0.03

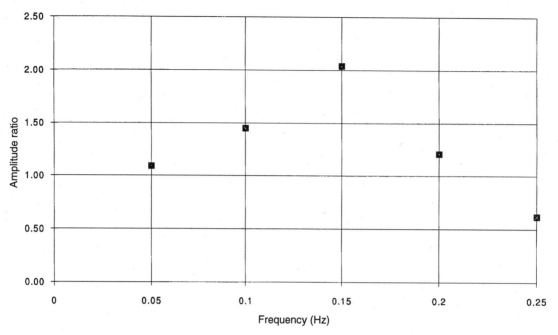

**Figure 6.16(a). Estimated amplitude ratio from numerical solution
of equation (6.6) math model.**

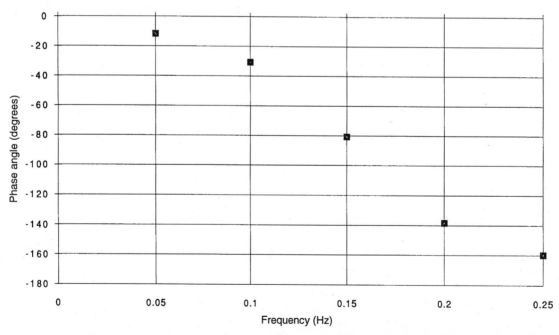

**Figure 6.16(b). Estimated phase lag from numerical solution
of equation (6.6) math model.**

coefficients are then determined so that both sides of the differential equation are satisfied for all time. We will shortly solve the second-order differential equation (6.6) in exactly this manner using a step function as the input.

Before we start investigating the exact solution, let's review *imaginary* and *complex* numbers. You've probably studied these numbers at some point in the past, but they may not have had much meaning, or may have sounded too "complex" and "imaginary" to be of much practical use. It turns out that complex numbers are *very* useful in engineering. They can provide you with additional tools that make life—or at least math model analysis—much easier. Having a working knowledge of complex numbers will allow you to solve differential equations faster and easier. So bear with me through the next section, and trust me that it will make modeling and analyzing engineering systems much easier.

Complex Numbers

You will recall from algebra that an attempt to find the solution to the equation

$$x^2 + 1 = 0 \qquad (6.16)$$

results in taking the square root of a negative number. That is, if you proceed to solve for x in the usual fashion, you would first isolate x on the left side of the equal sign

$$x^2 = -1$$

and then take the square root of both sides to find x

$$x = \pm\sqrt{-1}$$

The $\pm$ sign reminds you that there are two possible roots because of the square root sign.

Since we also learned that any number, positive or negative, times itself results in a positive number, then the square root of -1 is not allowed or is "undefined." To get around this problem, the square root of -1 was defined as an *imaginary number* and given the symbol "i." It has the property that $i \times i$ equals -1. Using this definition, we can easily confirm that $\pm i$ are the roots of (6.16) as follows:

$$(x - i)(x + i) = x^2 - ix + ix - i^2$$
$$= x^2 - (-1) = x^2 + 1$$

The number "i" is really no more "imaginary" than the number 1 is "real." Both real and imaginary numbers are simply words we use to describe a mental concept. So don't let names like *imaginary* and *complex* put you off. Incidentally, just so we don't confuse "i" with electrical current, I will use "j" to represent $\sqrt{-1}$ throughout the remainder of this book. This is common practice in engineering.

You will also recall from algebra that a quadratic equation is a polynomial of the form

$$ax^2 + bx + c = 0 \qquad (6.17)$$

For example, equation (6.16) is a quadratic equation in which $a = c = 1$ and $b = 0$. You will also recall that there are two roots, r_1 and r_2, to a quadratic equation such that

$$(x - r_1)(x - r_2) = ax^2 + bx + c = 0$$

where r_1 and r_2 are given by

$$r_1, r_2 = \frac{-b \pm \sqrt{b^2 - 4ac}}{2a}$$

or in the equivalent form by

$$r_1, r_2 = -\frac{b}{2a} \pm \sqrt{\left(\frac{b}{2a}\right)^2 - \frac{c}{a}} \qquad (6.18)$$

You can see that if $a = c = 1$ and $b = 0$ as with (6.16) then (6.18) gives

$$r_1, r_2 = -0 \pm \sqrt{0 - 1} = \pm\sqrt{-1}$$

which is the same result we obtained before.

If you look carefully at (6.18) you can see that as long as

$$\left(\frac{b}{2a}\right)^2 \geq \frac{c}{a}$$

the roots of the quadratic equation (6.17) will be real. But if

$$\left(\frac{b}{2a}\right)^2 < \frac{c}{a}$$

the roots will be complex. That is, they will contain both a real and an imaginary part. The roots could then be written as

$$r_1, r_2 = -\frac{b}{2a} \pm j \sqrt{\frac{c}{a} - \left(\frac{b}{2a}\right)^2}$$

or in a simpler form as

$$r_1, r_2 = -\sigma \pm j\omega \qquad (6.19)$$

where

$$\sigma = \frac{b}{2a}$$

and

$$\omega = \sqrt{\frac{c}{a} - \left(\frac{b}{2a}\right)^2}$$

The two roots expressed in the form of (6.19) are called *complex numbers* because they contain a *real* part σ and an *imaginary* part ω.

It's easy to visualize complex numbers graphically by creating a plane with two perpendicular axes. On one axis we can show real numbers and on the other axis imaginary numbers. Such a plane is called the *complex plane* or *imaginary plane* as it is used to plot complex numbers. We can plot the roots given by (6.19) on this plane as shown in Figure 6.17.

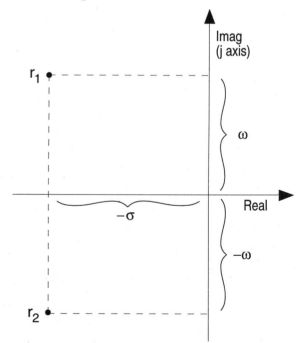

Figure 6.17. Roots of equation (6.19) plotted on the complex plane.

The two complex numbers given by (6.19) and shown in Figure 6.17 are said to be *complex conjugates* because they both have identical real parts and equal, but opposite in sign, imaginary parts. I'll use the symbol * to denote complex conjugates. Using this notation, we can rewrite the roots given in (6.19) as

$$r = \sigma + j\omega$$

$$r^* = \sigma - j\omega$$

If we draw a line from the origin to the points representing the location of the two roots as in Figure 6.18, then the plot looks like a polar plot and this suggests that we can express the two roots using polar coordinates, where each number is expressed in terms of its distance from the origin (an absolute value or magnitude) and an angle from the horizontal axis. Thus,

$$r_1 = |r_1| \angle \varphi = \sqrt{\sigma^2 + \omega^2} \, \angle \varphi$$

$$r_2 = |r_2| \angle -\varphi = \sqrt{\sigma^2 + \omega^2} \, \angle -\varphi$$

where

$$\varphi = \tan^{-1}\left(\frac{\omega}{\sigma}\right)$$

The polar notation used in Figure 6.18 also allows us to write a complex number in terms of trigonometric functions. The imaginary part of the complex number can be written as

$$\omega = |r_1| \sin \varphi \qquad (6.20)$$

and the real part as

$$\sigma = |r_1| \cos \varphi \qquad (6.21)$$

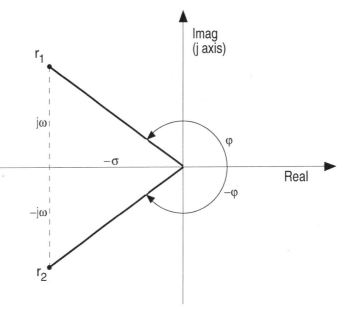

Figure 6.18. Plot of roots using polar notation.

From this you can see that

$$\sigma^2 + \omega^2 = r_1^2 \cos^2 \varphi + r_1^2 \sin \varphi$$
$$= r_1^2 \left(\cos^2 \varphi + \sin^2 \varphi\right)$$
$$= r_1^2 (1) = r_1^2$$

Using (6.20) and (6.21) we can rewrite the roots expressed in the form of (6.19) as

$$r_1 = r_1 \cos \varphi + jr_1 \sin \varphi$$
$$= |r_1|(\cos \varphi + j \sin \varphi)$$

and

$$r_2 = r_2 \cos \varphi - jr_2 \sin \varphi$$
$$= |r_2|(\cos \varphi - j \sin \varphi)$$

At this point you may be wondering why I have written the roots in this trigonometric form. I did this because I want to show you the following very important way to represent complex numbers:

$$|r_1|(\cos\varphi + j\sin\varphi) = r_1 e^{j\varphi} \qquad (6.22)$$

and

$$|r_2|(\cos\varphi - j\sin\varphi) = r_2 e^{-j\varphi} \qquad (6.23)$$

That is, two complex roots can be written as exponentials which can in turn be easily differentiated, integrated, multiplied, and divided.

To see this, recall from algebra and trigonometry that the cosine and sine of an angle can be written in the form of power series

$$\cos\varphi = 1 - \frac{\varphi^2}{2!} + \frac{\varphi^4}{4!} - \frac{\varphi^6}{6!} + \dots$$

$$\sin\varphi = \varphi - \frac{\varphi^3}{3!} + \frac{\varphi^5}{5!} - \frac{\varphi^7}{7!} + \dots$$

Now write the bracketed terms on the left side of (6.22) as

$$\cos\varphi + j\sin\varphi = 1 + (j\varphi) + \frac{(j\varphi)^2}{2!}$$
$$+ \frac{(j\varphi)^3}{3!} + \frac{(j\varphi)^4}{4!} + \dots \qquad (6.24)$$

and the bracketed term on the left side of (6.23) as

$$\cos\varphi - j\sin\varphi = 1 + (-j\varphi) + \frac{(-j\varphi)^2}{2!}$$
$$+ \frac{(-j\varphi)^3}{3!} + \frac{(-j\varphi)^4}{4!} + \dots \qquad (6.25)$$

Also recall from algebra that an exponential can be expressed as the following power series:

$$e^x = 1 + x + \frac{x^2}{2!} + \frac{x^3}{3!} + \frac{x^4}{4!} + \dots \qquad (6.26)$$

If you now compare (6.26) with (6.24) and (6.25) you can see that

$$e^{j\varphi} = \cos\varphi + j\sin\varphi \qquad (6.27)$$

and

$$e^{-j\varphi} = \cos\varphi - j\sin\varphi \qquad (6.28)$$

So we can indeed write the complex roots of a quadratic equation in the form

$$r_1 = |r_1| e^{j\varphi} \qquad (6.29)$$

$$r_2 = |r_2| e^{-j\varphi} \qquad (6.30)$$

You can also see that if

$$r = |r| e^{j\varphi}$$

then

$$r* = |r| e^{-j\varphi}$$

Being able to express complex roots in the form of exponentials makes our life easier. For example the product of two complex numbers

$$z = a + jb = |z| e^{j\varphi_1}$$

$$w = c + jd = |w| e^{j\varphi_2}$$

can be simply expressed as

$$z \times w = |z| \times |w| e^{j(\varphi_1 + \varphi_2)}$$

and the quotient of two complex numbers can be simply expressed as

$$\frac{z}{w} = \frac{|z|}{|w|} e^{j(\varphi_1 - \varphi_2)}$$

There are also two important *operators* used on complex number that extract the real or imaginary part of a complex number. These operators are designated *Re( )* and *Im( )*. For example, if

$$z = a + jb$$

then

$$Re(z) = a$$

and

$$Im(z) = b$$

There are several useful applications of these operators. For example, when we deal with the frequency response of systems we often encounter the sine and cosine of $(\omega t + \varphi)$. This can be represented as

$$Re\left[e^{j(\omega t + \varphi)} \right] = Re\left[\cos(\omega t + \varphi) + j\sin(\omega t + \varphi) \right]$$
$$= \cos(\omega t + \varphi)$$

and

$$Im\left[e^{j(\omega t + \varphi)} \right] = Im\left[\cos(\omega t + \varphi) + j\sin(\omega t + \varphi) \right]$$
$$= \sin(\omega t + \varphi)$$

We also saw in Chapter 4 that there is at least one root of the characteristic equation associated with a differential equation of the form

$$e^{r_1 t}$$

So when the roots are complex we can write

$$r_1 = \sigma + j\omega$$
$$e^{r_1 t} = e^{(\sigma t + j\omega t)} = e^{\sigma t} e^{j\omega t}$$

Then

$$Re\left(e^{r_1 t} \right) = Re\left(e^{\sigma t} e^{j\omega t} \right)$$
$$= Re\left(e^{\sigma t} \cos\omega t + j e^{\sigma t} \sin\omega t \right)$$
$$= e^{\sigma t} \cos\omega t$$

$$Im\left(e^{r_1 t} \right) = Im\left(e^{\sigma t} e^{j\omega t} \right)$$
$$= Im\left(e^{\sigma t} \cos\omega t + j e^{\sigma t} \sin\omega t \right)$$
$$= e^{\sigma t} \sin\omega t$$

Now that you are armed with a new set of tools, let's get back to finding the exact solution of a second-order linear ordinary differential equation with constant coefficients.

The Homogeneous Solution

Our problem is to find the exact solution to the math model for Figure 6.1 given by (6.6). We start by finding the homogeneous solution. That is, we first find the solutions to the equation

$$LC\frac{d^2 V_o}{dt^2} + RC\frac{dV_o}{dt} + V_o = 0 \qquad (6.31)$$

which is simply the left side of (6.6) set to zero.

We know that at least one solution to (6.31) exists in the form

$$V_o = A e^{rt}$$

Taking the derivatives of this solution

$$\frac{dV_o}{dt} = Are^{rt}$$

$$\frac{d^2V_o}{dt^2} = Ar^2e^{rt}$$

and substituting the solution and the derivatives into (6.31) gives

$$LC(Ar^2e^{rt}) + RC(Are^{rt}) + (Ae^{rt}) = 0 \quad (6.32)$$

Dividing both sides by

$$Ae^{rt}$$

gives

$$LCr^2 + RCr + 1 = 0 \quad (6.33)$$

You will immediately recognize equation (6.33) as a simple quadratic equation in r. This equation is often called the *characteristic polynomial* because it's derived from the characteristic equation. Compare (6.33) with (6.17) and you will see they have identical forms. The two roots r_1 and r_2 of the equation are given by (6.18). That is,

$$r_1 = -\frac{R}{2L} + \sqrt{\left(\frac{R}{2L}\right)^2 - \frac{1}{LC}} \quad (6.34)$$

and

$$r_2 = -\frac{R}{2L} - \sqrt{\left(\frac{R}{2L}\right)^2 - \frac{1}{LC}} \quad (6.35)$$

What all this means is that there are *two* solutions to (6.33), each of the form

$$Ae^{rt}$$

So the homogeneous solution is formed as a linear combination of these two solutions—that is,

$$(V_o)_H = A_1 e^{r_1 t} + A_2 e^{r_2 t} \quad (6.36)$$

where A_1 and A_2 are arbitrary constants to be determined from the initial conditions, and r_1 and r_2 are given by equations (6.34) and (6.35).

As you saw when we studied (6.18), if $(R/2L)^2$ is less than $(1/LC)$ in (6.34) and (6.35), a negative number exists under the square root sign and results in the roots being complex. In this case we can write the roots as

$$r_1 = -\left(\frac{R}{2L}\right) + j\sqrt{\left(\frac{1}{LC}\right) - \left(\frac{R}{2L}\right)^2}$$

or

$$r_1 = -\sigma + j\omega \quad (6.37)$$

and

$$r_2 = -\left(\frac{R}{2L}\right) - j\sqrt{\left(\frac{1}{LC}\right) - \left(\frac{R}{2L}\right)^2}$$

or

$$r_2 = -\sigma - j\omega = r_1{}^* \quad (6.38)$$

Equations (6.37) and (6.38) can be used to rewrite the homogeneous solution given by (6.36). That is,

$$(V_o)_H = A_1 e^{-\sigma t} e^{j\omega t} + A_2 e^{-\sigma t} e^{-j\omega t}$$

$$= e^{-\sigma t}\left(A_1 e^{j\omega t} + A_2 e^{-j\omega t}\right) \quad (6.39)$$

We can also express the homogeneous solution in terms of sines and cosines. We know that

$$e^{j\omega t} = \cos\omega t + j\sin\omega t \qquad (6.40)$$

and

$$e^{-j\omega t} = \cos\omega t - j\sin\omega t \qquad (6.41)$$

Substituting (6.40) and (6.41) into (6.39) gives

$$(V_o)_H = e^{-\sigma t}(A_1\cos\omega t + jA_1\sin\omega t$$
$$+ A_2\cos\omega t - jA_2\sin\omega t)$$

$$= \sigma^{-\sigma t}\left[A_1 + A_2\right)\cos\omega t$$
$$+ j(A_1 - A_2)\sin\omega t \;\right] \quad (6.42)$$

Let's introduce two new arbitrary constants, B_1 and B_2, and require them to be real. That is, we let

$$B_1 = A_1 + A_2 \qquad (6.43)$$

$$B_2 = j(A_1 - A_2) \qquad (6.44)$$

Substituting (6.43) and (6.44) into (6.42) gives the homogeneous solution in the form

$$(V_o)_H = \sigma^{-\sigma t}\left[B_1\cos\omega t + B_2\sin\omega t\right] \quad (6.45)$$

I want you to note that requiring B_1 and B_2 to be real means that A_1 and A_2 are arbitrary complex conjugate constants. You see this by solving (6.43) and (6.44) for A_1 and A_2. That is,

$$A_1 = \frac{1}{2}B_1 - j\frac{1}{2}B_2 = A \qquad (6.46)$$

$$A_2 = \frac{1}{2}B_1 + j\frac{1}{2}B_2 = A* \qquad (6.47)$$

What this means is the homogeneous solution given in the form of (6.45) can also be written in the equivalent form

$$(V_o)_H = A_1e^{-rt} + A*e^{-r*t}$$

where the * denotes complex conjugates.

There are two other cases associated with equations (6.34) and (6.35) that must be investigated. A special case occurs when

$$\left(\frac{R}{2L}\right)^2 = \left(\frac{1}{LC}\right)$$

In this case the term under the square root sign is zero and both roots are equal and real. The other case occurs when

$$\left(\frac{R}{2L}\right)^2 > \left(\frac{1}{LC}\right)$$

In this case, the roots r_1 and r_2 are unequal and real.

For the case of equal (repeated) roots the homogeneous solution is formed by a linear combination of the roots as before, except one root gets multiplied by the independent variable t. That is the solution is of the form

$$(V_o)_H = A_1e^{-\sigma t} + A_2te^{-\sigma t} \qquad (6.48)$$

For the case of unequal real roots, the homogeneous solution is of the same form as when the roots were complex. That is

$$(V_o)_H = A_1e^{r_1t} + A_2e^{r_2t} \qquad (6.49)$$

Let's summarize by writing the homogeneous solutions for the three cases:

Case I - Real and unequal roots
(sometimes called *over damped*)

$$\left(\frac{R}{2L}\right)^2 > \left(\frac{1}{LC}\right)$$

$$\left(V_o\right)_H = A_1 e^{r_1 t} + A_2 e^{r_2 t} \qquad (6.49)$$
repeated

Case II - Real and equal roots
(sometimes called *critically damped*)

$$\left(\frac{R}{2L}\right)^2 = \left(\frac{1}{LC}\right)$$

$$\left(V_o\right)_H = A_1 e^{-\sigma t} + A_2 t e^{-\sigma t} \qquad (6.48)$$
repeated

Case III - Imaginary and unequal roots
(sometimes called *under damped* or *oscillatory*)

$$\left(\frac{R}{2L}\right)^2 < \left(\frac{1}{LC}\right)$$

$$\left(V_o\right)_H = e^{-\sigma t}\left(B_1 \cos \omega t + B_2 \sin \omega t\right)$$
(6.45)
repeated

Before we move on to investigate the particular solution to a second-order differential equation, I want you to look at a graphical representation of the roots to the characteristic polynomial. We can represent all possible roots of the characteristic equation by plotting them on the complex plane. Let's start with Case I and consider first the condition when

$$\left(\frac{R}{2L}\right)^2 >> \left(\frac{1}{LC}\right)$$

This could occur, for instance, when the capacitor in the circuit is very large. Equation (6.34) for the first root then becomes

$$r_1 = -\frac{R}{2L} + \sqrt{\left(\frac{R}{2L}\right)^2 - (\approx 0)}$$

$$= -\frac{R}{2L} + \frac{R}{2L} = 0$$

and equation (6.35) for the second root becomes

$$r_2 = -\frac{R}{2L} - \sqrt{\left(\frac{R}{2L}\right)^2 - (\approx 0)}$$

$$= -\frac{R}{2L} - \frac{R}{2L} = -\frac{R}{L}$$

We can plot these two roots on the complex plane as shown in Figure 6.19.

Now look at Case II. The roots are equal and given by

$$r_1 = r_2 = -\frac{R}{2L}$$

We plot these on the complex plane and note that the two roots are now located along the negative real axis exactly one-half the distance between the starting points for the Case I roots. Of course there are an infinite number of roots between the Case I extreme (where the capacitor is extremely large) and the Case II condition. The heavy solid lines with arrows represent the *loci* of all of these roots. If for example you

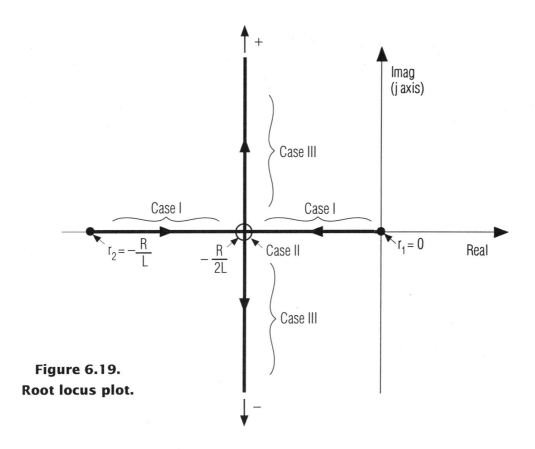

Figure 6.19.
Root locus plot.

held R and L constant and varied C from infinity toward zero, you could compute all of the roots to the characteristic equation. All Case I roots would lie along the negative real axis and the Case II roots would lie on the real axis at $-R / 2L$.

We can now plot the Case III roots using the same concept of ever-decreasing value for the capacitor while holding the resistor and inductor values constant. The Case III roots are

$$r_1 = -\left(\frac{R}{2L}\right) + j\sqrt{\left(\frac{1}{LC}\right) - \left(\frac{R}{2L}\right)^2}$$

and

$$r_2 = -\left(\frac{R}{2L}\right) - j\sqrt{\left(\frac{1}{LC}\right) - \left(\frac{R}{2L}\right)^2}$$

You can see that as C decreases, $(1 / LC)$ increases, but $(R / 2L)$ remains constant. The real parts of the roots stay constant while the imaginary parts increase toward $\pm$ infinity.

The plot shown in Figure 6.19 is called a *root locus plot*. It is very handy for visualizing the behavior of a system as parameters are changed. It is also used for studying the stability of engineering systems, a subject we will discuss later.

The Solution for a Step Input

Now that we have the homogeneous solution to equation (6.6), all we need now is the particular solution. You will recall from Chapter 4 that the method of undetermined coefficients consists of constructing a particular solution $(V_o)_P$ that looks similar to the input or forcing function but has unknown coefficients. The coefficients are determined by substituting the particular solution into the differential equation and finding values for the coefficients that make the equation hold for all time.

For a step input function, the input voltage suddenly rises to a value V_{is} and remains at this level for all time. A particular solution that looks like this input function is simply a constant, k_o. That is

$$(V_o)_P = k_o$$

When this solution is substituted in (6.6), all of the derivatives of the particular solution are zero and we are left with

$$k_o = V_{is}$$

which satisfies (6.6) for all time. The particular solution for a step input is thus given by

$$(V_o)_P = V_{is} \qquad (6.50)$$

We can now summarize the complete solution of (6.6) for a step input function.

Case I - Real and unequal roots
(sometimes called *over damped*)

$$\left(\frac{R}{2L}\right)^2 > \left(\frac{1}{LC}\right)$$

$$V_o = V_{is} + A_1 e^{r_1 t} + A_2 e^{r_2 t} \qquad (6.51)$$

Case II - Real and equal roots
(sometimes called *critically damped*)

$$\left(\frac{R}{2L}\right)^2 = \left(\frac{1}{LC}\right)$$

$$V_o = V_{is} + A_1 e^{-\sigma t} + A_2 t e^{-\sigma t} \qquad (6.52)$$

Case III - Imaginary and unequal roots
(sometimes called *under damped* or *oscillatory*)

$$\left(\frac{R}{2L}\right)^2 < \left(\frac{1}{LC}\right)$$

$$V_o = V_{is} + e^{-\sigma t}\left(B_1 \cos \omega t + B_2 \sin \omega t\right) \qquad (6.53)$$

The coefficients A_1 and A_2, or B_1 and B_2, must be determined using the initial conditions $(V_o)_{init}$ and $(dV_o/dt)_{init}$. If both of these conditions are taken to be zero—i. e., the system is initially at rest so there is no charge on the capacitor or no current—then the coefficients can be determined as follows:

For Case I

At time t = 0, all exponentials of the form e^{rt} are equal to 1. We can therefore write the equations for the initial conditions as

$$(V_o)_{init} = 0 = V_{is} + A_1 + A_2$$

$$(\dot{V}_o)_{init} = 0 = 0 + r_1 A_1 + r_2 A_2$$

This provides two simultaneous equations for A_1 and A_2 which can easily be solved giving

$$A_1 = \frac{r_2}{r_1 - r_2} V_{is}$$

$$A_2 = -\frac{r_1}{r_1 - r_2} V_{is}$$

Substituting into (6.51) gives

$$V_o = V_{is} \left[1 + \left(\frac{r_2}{r_1 - r_2} \right) e^{r_1 t} - \left(\frac{r_1}{r_1 - r_2} \right) e^{r_2 t} \right]$$

$$(6.54)$$

For Case II

We proceed the same way and write the initial condition equations as

$$\left(V_o \right)_{init} = 0 = V_{is} + A_1$$

$$\left(\dot{V}_o \right)_{init} = 0 = 0 + r_1 A_1 + A_2$$

These are readily solved giving

$$A_1 = -V_{is}$$

$$A_2 = r_1 V_{is}$$

Substituting into (6.52) gives

$$V_o = V_{is} \left[1 + e^{r_1 t} \left(r_1 t - 1 \right) \right] \qquad (6.55)$$

For Case III

We again proceed as above and write the initial condition equations as

$$\left(V_o \right)_{init} = 0 = V_{is} + B_1$$

$$\left(\dot{V}_o \right)_{init} = 0 = 0 + \omega B_2 - \sigma B_1$$

These are readily solved giving

$$B_1 = -V_{is}$$

$$B_2 = -\frac{\sigma}{\omega} V_{is}$$

Substituting into (6.53) gives

$$V_o = V_{is} \left[1 - \sqrt{1 + \left(\frac{\sigma}{\omega} \right)^2} \, e^{-\sigma t} \sin \left(\omega t + \phi \right) \right] \quad (6.56)$$

where

$$\phi = \tan^{-1} \frac{\omega}{\sigma}$$

We are now at a place where we can compare the exact solutions given here with the numerical solutions we obtained earlier. However, before we do this you should get used to looking at a second-order linear ordinary differential equation with constant coefficients in the following form:

$$\frac{1}{\omega_n^2} \frac{d^2 V_o}{dt^2} + \frac{2\xi}{\omega_n} \frac{dV_o}{dt} + V_o = V_i \qquad (6.57)$$

where ω_n = the *natural frequency* of the system

ζ = the *damping ratio* of the system.

Compare equation (6.6) with (6.57). You will see that

$$\frac{1}{\omega_n^2} = LC \Rightarrow \omega_n = \sqrt{\frac{1}{LC}} \qquad (6.58)$$

and

$$\frac{2\xi}{\omega_n} = RC \Rightarrow \xi = \frac{R}{2} \sqrt{\frac{C}{L}} \qquad (6.59)$$

We can rewrite the roots of the characteristic equation using the natural frequency and the damping ratio as follows

$$r_1, r_2 = -\frac{b}{2a} \pm \sqrt{\left(\frac{b}{2a}\right)^2 - \frac{c}{a}}$$

$$r_1, r_2 = -\frac{1}{2}\frac{2\xi}{\omega_n}\omega_n^2 \pm \sqrt{\left(\frac{1}{2}\frac{2\xi}{\omega_n}\omega_n^2\right)^2 - \omega_n^2}$$

$$= -\xi\omega_n \pm \omega_n\sqrt{\xi^2 - 1} \qquad (6.60)$$

You can see that when $\zeta^2 < 1$,

$$r_1 = -\xi\omega_n + j\omega_n\sqrt{1-\xi^2} = r$$

and

$$r_2 = -\xi\omega_n - j\omega_n\sqrt{1-\xi^2} = r*$$

The equations are now identical to equations (6.37) and (6.38) and you can see from a comparison that

$$\sigma = \xi\omega_n \qquad (6.61)$$

and

$$\omega = \omega_n\sqrt{1-\xi^2} \qquad (6.62)$$

Substituting (6.61) and (6.62) into (6.56) gives

$$V_o = V_{is}\left[1 - \frac{1}{\sqrt{1-\xi^2}}e^{-\xi\omega_n t}\sin\left(\sqrt{1-\xi^2}\,\omega_n t + \phi\right)\right]$$

$$(6.63)$$

where

$$\phi = \tan^{-1}\left(\sqrt{\frac{1-\xi^2}{\xi^2}}\right)$$

We'll use equation (6.63) to study the effect of V_{is}, ζ, and ω_n on the step response of a second-order differential equation, just as we studied the effect of V_{is} and τ on the step response of a first-order differential equation in Chapter 4. At this time, you should write a program that will accept V_{is}, R, L and C and compute ζ, ω_n and $V_o(t)$ using equations (6.58), 6.59), and (6.63) respectively. Solve equation (6.63) from t = 0 to t = 4 / ω_n using 0.05 / ω_n steps for the following values

$$\begin{aligned}
L &= 1 \text{ henry} \\
C &= 1 \text{ farad} \\
R &= 0.6 \text{ ohm} \\
V_{is} &= 10 \text{ volts}
\end{aligned}$$

These are the same values you used for the first numerical solution, so you can compare your results. My comparative results are shown in Figure 6.20.

You should use the exact solution to solve for the step response for all the cases we studied in Section 6.3. I also suggest that you solve the equation for several values of ζ and ω_n to gain a feel for how these two parameters affect the step response. Note that you can show your results in a more general way by dividing both sides of (6.63) by V_{is}. That will make the equation dimensionless. Also note that $\omega_n t$ can be thought of as dimensionless time. So if you plot V_o / V_{is} versus $\omega_n t$ you will have a dimensionless step response to a second-order differential equation with ζ as a parameter. My version of this graph is shown in Figure 6.21.

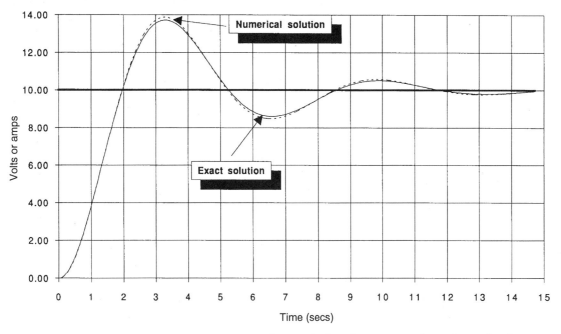

Figure 6.20. Step response of circuit in Figure 6.1 (R = 0.6 ohm).

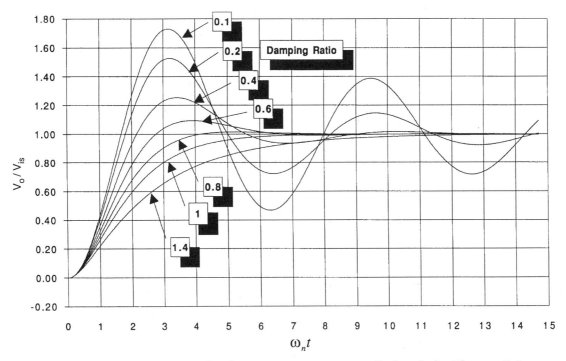

Figure 6.21. Dimensionless step response of circuit in Figure 6.1.

The Solution for a Sinusoidal Input

You already have a good insight into the frequency response of a second-order system from the numerical work in Section 6.4. You also realize that most of the time you will not be interested in the transient start-up that goes along with a suddenly applied sinusoidal forcing function. Instead, you will be mostly interested in the frequency response of the system. That is, determining how the output-to-input amplitude ratio and phase lag behave as a function of the forcing frequency.

I'll now show you a quick and easy way to obtain the frequency response. The method is applicable to any order system, but for now I will apply it to our math model cast in the form of equation (6.57). That is, we will find the nontransient solution of

$$\frac{1}{\omega_n^2}\frac{d^2V_o}{dt^2} + \frac{2\xi}{\omega_n}\frac{dV_o}{dt} + V_o = V_{is}\sin\omega_f t \quad (6.64)$$

for all forcing frequencies ω_f. The nontransient solution of this equation, you will recall, is the particular solution. The transient solution is the homogeneous solution. We are going to ignore this part.

We first take the case where the forcing function, i.e., the right side of (6.64), is a complex exponential

$$V_i = V_f e^{st} \quad (6.65)$$

where s and V_f can be complex numbers. Next assume that the particular solution is

$$V_o = V_r e^{st} \quad (6.66)$$

where V_r can also be a complex number.

Differentiate (6.66) twice

$$\frac{dV_o}{dt} = sV_r e^{st} \quad (6.67)$$

$$\frac{d^2V_0}{dt^2} = s^2V_r e^{st} \quad (6.68)$$

and substitute (6.65), (6.66), (6.67) and (6.68) into (6.64) and get

$$\frac{s^2}{\omega_n^2}V_r e^{st} + 2\xi\frac{s}{\omega_n}V_r e^{st} + V_r e^{st} = V_f e^{st} \quad (6.69)$$

The e^{st} terms can be cancelled out of both sides of the equation and the V_r terms can be factored out of the left side of the equation. This gives

$$\left(\frac{s^2}{\omega_n^2} + 2\xi\frac{s}{\omega_n} + 1\right)V_r = V_f$$

or

$$\frac{V_r}{V_f} = \frac{1}{\left(\frac{s^2}{\omega_n^2} + 2\xi\frac{s}{\omega_n} + 1\right)} \quad (6.70)$$

You can see that the ratio of the output to the input is simply the transfer function with the D operator replaced by the complex number s.

Recall from the discussion of complex numbers that the sinusoidal forcing function

$$V_i = V_{is}\sin\omega_f t$$

can be written as

$$V_i = Im\left[V_{is}e^{jw_f t}\right] \quad (6.71)$$

because

$$Im\left[V_{is}e^{jw_ft}\right] = Im\left[V_{is}\left(\cos\omega_ft + j\sin\omega_ft\right)\right]$$

$$= V_{is}\sin\omega_ft$$

You can see that (6.71) is a special case of the more general complex exponential function given by (6.65). In (6.71) $s = j\omega_f$ and V_i and V_{is} are real. So all you need to do is substitute $j\omega_f$ for s in the transfer function given by (6.66) and associate V_o with V_r and V_{is} with V_f to get the frequency response. That is

$$\frac{V_o}{V_{is}} = \frac{1}{\left(\dfrac{\left(j\omega_f\right)^2}{\omega_n^2} + 2\xi\dfrac{\left(j\omega_f\right)}{\omega_n} + 1\right)} \qquad (6.72)$$

Since $j2 = -1$, (6.72) simplifies to

$$\frac{V_o}{V_{is}} = \frac{1}{1 - \left(\dfrac{\omega_f}{\omega_n}\right)^2 + j2\xi\left(\dfrac{\omega_f}{\omega_n}\right)} \qquad (6.73)$$

Note that the ratio of the amplitude of the output and the amplitude of the input is a complex number. Since any complex number can be written as a magnitude and an angle, (6.73) can also be written as

$$\frac{V_o}{V_{is}} = \frac{1}{\sqrt{\left[1 - \left(\dfrac{\omega_f}{\omega_n}\right)^2\right]^2 + \left[2\xi\left(\dfrac{\omega_f}{\omega_n}\right)\right]^2}}e^{-j\phi} \qquad (6.74)$$

where
$$\phi = \tan^{-1}\left[-\frac{2\xi\left(\dfrac{\omega_f}{\omega_n}\right)}{1 - \left(\dfrac{\omega_f}{\omega_n}\right)^2}\right] \qquad (6.75)$$

Before we look at plots of equations (6.74) and (6.75), I would like you to review Table 6.8, which provides the step-by-step procedure for obtaining the frequency response of any linear, ordinary differential equation with constant coefficients. I strongly recommend you commit this procedure to memory. It will work for *any* order system.

Table 6.8. Simple way to obtain the frequency response.

Step 1 - Put the system differential equation in operator transfer function form.

Example:
$$\tau\frac{dV_o}{dt} + V_o = V_i$$
$$(\tau D + 1)V_o = V_i$$
$$\frac{V_o}{V_i} = \frac{1}{\tau D + 1}$$

Step 2 - Substitute $j\omega_f$ for the operator D.
$$\frac{V_o}{V_i} = \frac{1}{j\tau\omega_f + 1}$$

Step 3 - Express resulting complex number as a magnitude and an angle.
$$\frac{V_o}{V_i} = \frac{1}{\sqrt{1 + \tau^2\omega_f^2}}e^{-j\phi}$$
$$\phi = \tan^{-1}\left(-\tau\omega_f\right)$$

NOTE: *Compare this method to the one used in Chapter 4.*

Figure 6.22 shows a plot of (6.74) and (6.75). These two plots are the frequency response of our system for various values of the damping ratio. Carefully study this figure and the associated frequency response plots. You are likely to encounter the frequency response of second-order systems countless times in your technical career, so there are a number of important points to note.

First note in (6.74) that when the forcing frequency ω_f is equal to the natural frequency ω_n of the system, the term

$$\left[1-\left(\frac{\omega_f}{\omega_n}\right)^2\right]$$

becomes zero, the amplitude ratio becomes

$$\frac{V_o}{V_{is}} = \frac{1}{2\xi} \qquad (6.76)$$

and the phase lag equals

$$\phi = \tan^{-1}\left[-\frac{2\xi}{0}\right] = \tan^{-1}(-\infty) = -90°$$

Note in (6.76) that the amplitude ratio at this critical frequency is controlled solely by the damping ratio. For small values of damping, the amplitude ratio can become quite large. This amplification of the input has its good and bad sides. It is often used in electrical circuits to create oscillators. Vibration absorbers sometimes found in mechanical systems also make use of this phenomena. However, in many mechanical systems this phenomena can amplify stresses in structural members that can lead to their failure. Earthquakes and strong winds can create forcing functions that have frequencies

near the natural frequency of a structure. These can excite the structure at its natural frequencies, and if the structure cannot withstand the amplified motions and stresses, it will collapse.

The second point to note is that when ω_f is small compared to ω_n, the amplitude of the output sinusoid is approximately equal to the amplitude of the input sinusoid. The frequency response is said to be *flat* in this region. On the other hand, when ω_f is large compared to ω_n, the amplitude ratio decreases in proportion to $(\omega_f / \omega_n)^{-2}$. You can see this by looking at only the magnitude of the frequency response given by (6.74). That is,

$$\left|\frac{V_o}{V_{is}}\right| = \frac{1}{\sqrt{\left[1-\left(\frac{\omega_f}{\omega_n}\right)^2\right]^2 + \left[2\xi\left(\frac{\omega_f}{\omega_n}\right)\right]^2}}$$

Expanding terms under the square root sign gives

$$\left|\frac{V_o}{V_{is}}\right| = \frac{1}{\sqrt{1-2\left(\frac{\omega_f}{\omega_n}\right)^2 + \left(\frac{\omega_f}{\omega_n}\right)^4 + 4\xi^2\left(\frac{\omega_f}{\omega_n}\right)^2}}$$

When $\omega_f \gg \omega_n$, then the $(\omega_f / \omega_n)^4$ term dominates so

$$\left|\frac{V_o}{V_{is}}\right| \approx \frac{1}{\sqrt{\left(\frac{\omega_f}{\omega_n}\right)^4}} = \frac{1}{\left(\frac{\omega_f}{\omega_n}\right)^2} = \left(\frac{\omega_f}{\omega_n}\right)^{-2}$$

In essence, the amplitude ratio decreases by a factor of 100 (−40 db) for every decade increase in the frequency when $\omega_f \gg \omega_n$.

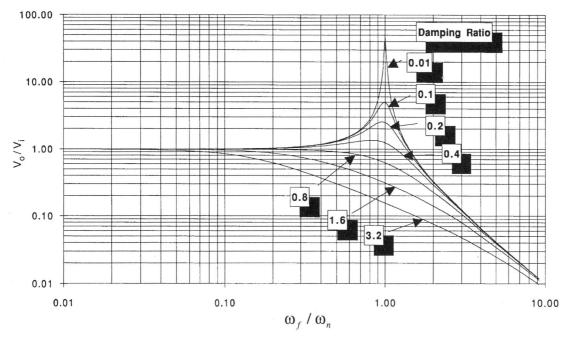

Figure 6.22(a). Frequency response (amplitude ratio) of second-order math model [see equation (6.74)].

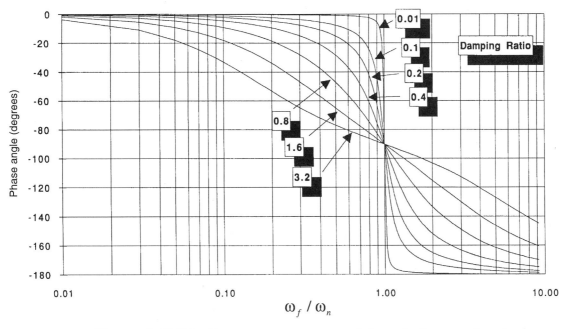

Figure 6.22(b). Frequency response (phase) of second-order math model [see equation (6.75)].

In Chapter 4 you studied the low-pass filtering characteristics of the circuit presently being studied without the inductor. If this circuit were actually being used to filter unwanted frequencies from an input signal, you might want to compare the frequency response characteristics of the two circuits to see which is the better low-pass filter. From an ideal low-pass signal filter standpoint, all signals within the passband of the filter should pass without any attenuation and without any phase shift. When the input frequency goes above the cutoff frequency, the output signal should be completely attenuated; i. e., the amplitude ratio should be zero.

If we select the cutoff frequency for the circuit of Figure 6.1 to be ω_n, then the amplitude ratio is given by equation (6.74). Selecting $\zeta = 0.707$ gives an amplitude ratio of 0.707 or –3 db. This amplitude ratio is then the same as it is for the circuit of Figure 4.1 when the cutoff frequency was $1/\tau$. Figure 6.23 shows a comparison of the two circuits with each other and the ideal filter. Clearly the circuit given in Figure 6.1 is better than the circuit given in Figure 4.1 insofar as the amplitude ratio is concerned. The second-order circuit is better able to pass signals

without attenuation within the passband and is better able to attenuate unwanted signals in the stop band. But, as you can see, there is a penalty paid for the better amplitude performance of the second-order system when we look at phase shift. The second-order system causes a greater phase shift between input and output signals than does the first-order system. Filters are often used in data acquisition systems to remove unwanted, high-frequency noise that can cause aliasing errors in the digitization process. Phase shift between input and output signals caused by filters can become a serious problem when two channels are being compared on a common time base and each uses filters with different characteristics.

Now let's summarize our analysis of the second-order electrical system we've been studying. Figure 6.24 provides the summary in the same form used for the first-order electrical system studied in Chapter 4. Compare this figure with Figure 4.18. For a first-order system, only one parameter, τ, was needed to characterize the system. For a second-order system, two parameters, ζ and ω_n, are required to completely characterize the system.

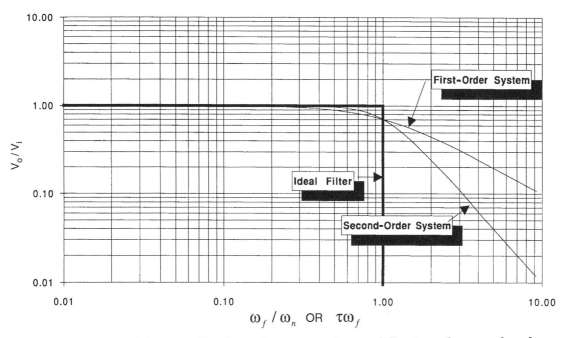

Figure 6.23(a). Amplitude ratio comparison of first- and second-order systems with an ideal filter.

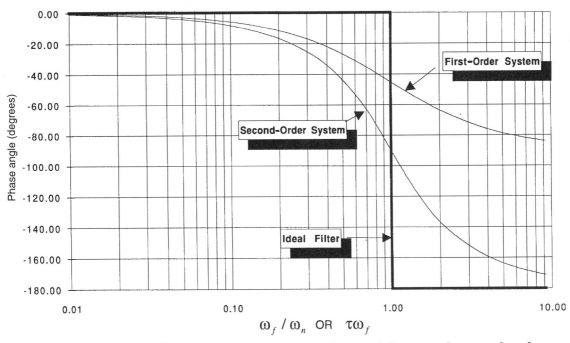

Figure 6.23(b). Phase angle comparison of first- and second-order systems with an ideal filter.

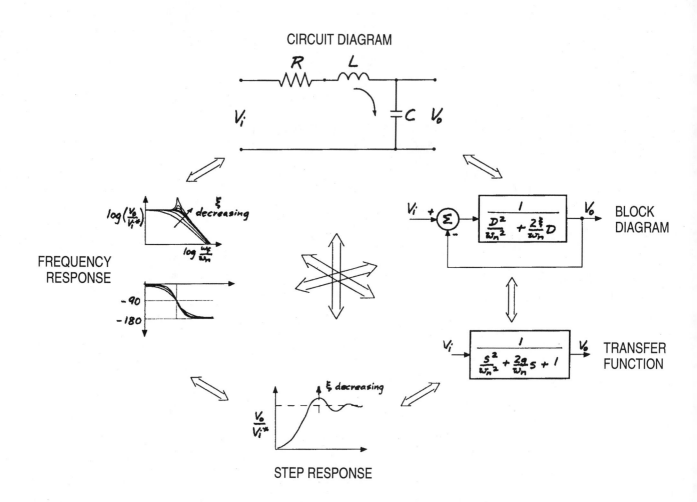

**Figure 6.24. Summary of second-order
electrical system.**

Chapter

7 Practical Applications of Second-Order Math Models

Objectives

When you have completed this chapter, you will be able to:

- ■ Construct a math model of an electro-hydraulic-mechanical engineering system using any of the four tools presented in Chapter 4.

- ■ Use the tools you learned in earlier chapters to size system components.

- ■ Use the advanced tools you learned in Chapter 6 to analyze the system.

- ■ Reduce higher-order math models to second- or first-order approximation models.

7.1 Introduction

In this chapter we will apply model construction and analysis techniques to a real-life engineering system problem involving the design of a laboratory water wave-making machine. I have selected this example because the system employs translational mechanical, hydraulic, and electrical elements all in one system. (And it happens to be one that I'm intimately familiar with!) It will reinforce the important concepts of analogies and system models that look alike mathematically.

You will learn how to build models of hydraulic actuators and valves, and how to combine these with electrical and mechanical models. In general, hydraulic power system models are nonlinear. You'll learn how to apply a nonlinear hydraulic model to the steady-state design problem, and then how to linearize the model and use it to study the dynamic behavior of the system.

This chapter also introduces you to the concepts of automatic control and feedback systems. Such systems are really no more complicated or different than the systems you have studied so far. With this introduction to automatic controls, you should be able to understand the fundamentals of such systems.

7.2 Electro-Hydraulic-Mechanical System Case Study
Problem Definition

Figure 7.1 shows a sketch of a device used to generate water waves in a laboratory. It consists of a water-filled flume containing a hinged flap at one end and a wave absorption beach at the other. As the flap is moved back and forth, waves are created on the water surface that travel from the flap toward the beach. Such machines are used by coastal and ocean engineers to study the behavior of ocean waves and the forces they exert on marine structures.

We are going to design an actuator (which simply means a device for moving or controlling something indirectly instead of by hand) that will move the wave flap, so it can generate a wave having an amplitude up to 1.5 feet at a period of around 2.4 seconds. Let's presume the ocean engineers involved with the project have

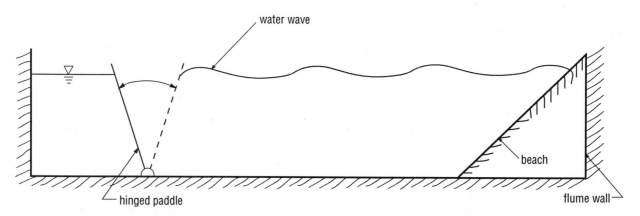

water wave

beach

hinged paddle

flume wall

Figure 7.1. System for generating water waves in a laboratory.

already performed experiments and analyzed the forces that the water exerts on the flap. For sinusoidal flap motion of the form

$$S = S_o \sin\left(\frac{2\pi}{T}t\right)$$

where S is the linear displacement of the top of the flap and T is the period, their load calculations are summarized in Table 7.1.

The force F_R is a sinusoidal hydrodynamic force that is *in phase with the stroke velocity* and the force F_I is a sinusoidal hydrodynamic and wave flap inertial force that is *in phase with the stroke acceleration*. The values for F_R and F_I given are peak *amplitude* values.

The ocean engineers want to control the wavemaker with a computer. They want our system to be able to accept a voltage output from their computer. The amplitude of the voltage signal will control the amplitude of the wave flap, and the frequency will control the period of the flap. They also want us to tell them what the

transfer function between the input voltage and the output wave flap stroke will be after the system is built. They plan to use the transfer function to compensate for any discrepancies between the wave height commanded by the computer and what they actually measure in the flume.

Conceptual Design Considerations

The way the problem has been stated so far, there are a number of actuator systems that could be used to produce sinusoidal motion of the flap under control of a voltage signal from a computer. These might include a DC or AC motor driving a lead screw attached to the wave flap, a DC or AC motor driving a set of pulleys and cables attached to the wave flap, or a hydraulic valve and ram attached to the wave flap. Given complete freedom of choice and the time to investigate all alternatives, we should try to design the actuator system to fit within the space available and to have the lowest life cycle cost. That

Table 7.1.
Load Calculation Summary.

Condition	Period, T (seconds)	Maximum Stroke, S_o (inches)	Resistive Force, F_R (pounds)	Inertial Force, F_I (pounds)
1	0.55	0.787	2.9	397.9
2	1.00	3.386	36.0	359.7
3	2.00	13.779	323.7	29.2
4	2.40	25.984	616.0	15.7
5	2.90	24.409	483.4	36.0
6	4.00	20.472	330.5	38.2
7	5.00	19.685	260.8	29.2

is, we should seek the alternative that minimizes the capital, operating, and maintenance costs over the desired life of the system. The time we have to solve the problem, as well as our experience with various system alternatives, will also play a role in our design approach.

In this chapter we will restrict our attention to the design of a single concept employing an electrically actuated hydraulic valve connected to a hydraulic actuator. The concept is shown in Figure 7.2.

The actuator will be mounted on the end wall of the flume. One end of the ram will be attached to the top of the wave flap. The hydraulic power supply will be mounted near the actuator at the flap end of the flume. We will only look at the technical side of the problem and not investigate capital, operating, and maintenance costs. In practice, you would likely have to obtain capital costs of equipment from vendors and estimate installation, operating, and maintenance costs.

Derivation of Steady-state Math Model of System

Let's begin the development of this math model with Figure 7.3. This is a schematic diagram of a *hydraulic actuator* (often called a *ram*) and a so-called *four-way hydraulic valve*. The purpose of the actuator is to develop a large force at a relatively high velocity using a high pressure hydraulic fluid. The purpose of the hydraulic valve is to control the quantity and direction of the hydraulic fluid flowing into and out of the actuator.

We'll first create a steady-state math model for the hydraulic valve, and then one for the actuator. Then we'll combine them to get a steady-state equation for the entire system.

Steady-state Model for the Hydraulic Valve

The hydraulic valve is much smaller than the actuator and its ports are often connected to

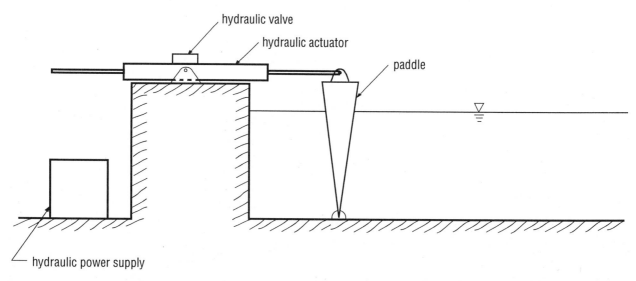

Figure 7.2. Hydraulic system for generating waves.

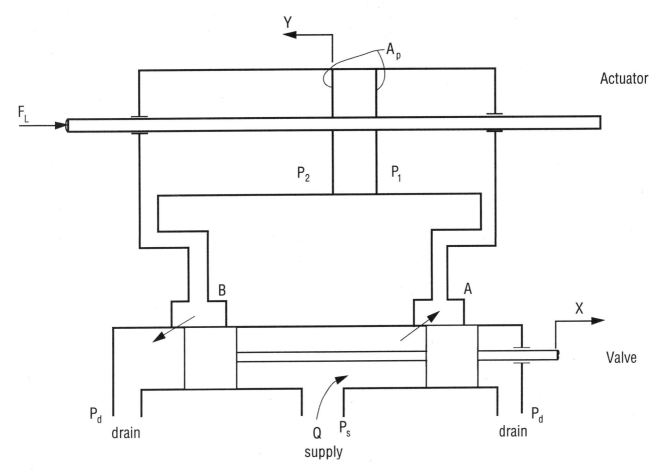

Figure 7.3. Definition sketch of hydraulic valve (four-way) and hydraulic actuator (double-acting).

the ram using hydraulic hoses and to the hydraulic power supply with steel pipe. Quite often the position of the hydraulic valve is controlled by an electric actuator. We will consider such a valve in a moment, but for the time being consider the valve being positioned by a lever or by hand.

We want a math model that relates the output position of the ram, Y, to the input position of the valve, X. Figure 7.4 shows a close-up of Port A of the valve.

As a first approximation, we will assume the flow from the valve chamber is steady and apply Bernoulli's Equation (see Chapter 2) along streamlines that exist between the supply chamber of the valve and the exit port. That gives

$$P_s + \frac{\rho}{2}V_s^2 = P_1 + \frac{\rho}{2}V_1^2 \qquad (7.1)$$

We can assume that the velocity of the supply fluid inside the chamber is much smaller than the velocity of the fluid when it exits the chamber. So we rewrite (7.1) as

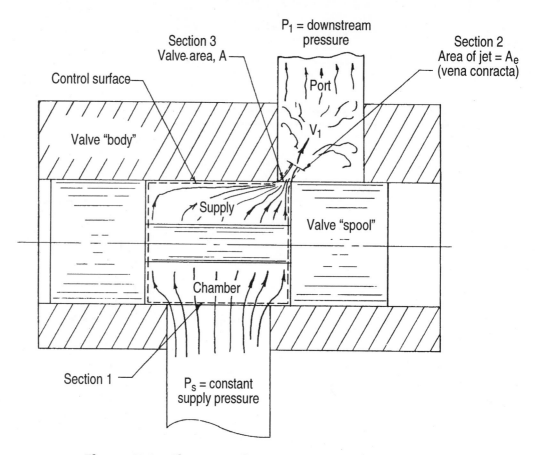

Figure 7.4. Close-up of Port A of the valve.

$$\frac{\rho}{2}V_1^2 = P_s - P_1$$

or

$$V_1 = \sqrt{\frac{2}{\rho}\left(P_s - P_1\right)} \qquad (7.2)$$

The volume of hydraulic fluid flowing from the supply chamber into the port is equal to the product of the effective area of the port opening and the velocity. That is

$$Q = V_1 A_e \qquad (7.3)$$

The geometric area of the opening is equal to

the product of the circumference of the valve spool, W, and the width of the opening, X. The effective area of the opening is not quite equal to the geometric area. When dealing with sharp-edged orifices, it has been found that the effective area through which the fluid flows is smaller than the actual geometric area. The ratio of these two areas is called the discharge coefficient, C_d, and the effective area is expressed as

$$A_e = C_d W X \qquad (7.4)$$

The value of C_d has been determined experimentally by others and found to equal 0.625.

Combining (7.2), (7.3) and (7.4) gives the flow through this single port as

$$Q = C_d W X \sqrt{\frac{2}{\rho}(P_s - P_1)} \qquad (7.5)$$

This is an important equation when working with fluid power systems. Think of the equation as a math model for a variable, nonlinear fluid resistor. Later we will linearize this equation when we proceed to develop our linear system model, but for now we will leave it in its non-linear form.

Note in (7.5) that C_d, W, and ρ are constants. We can combine all constant terms into one valve constant, C_v, and simplify the equation to

$$Q = C_v X \sqrt{P_s - P_1} \qquad (7.6)$$

where

$$C_v = C_d W \sqrt{\frac{2}{\rho}} \qquad (7.7)$$

Notice in (7.6) that when the valve is at it maximum opening, X_{max}, and when P_1 is zero, the flow through Port A is at its maximum value. We can write this maximum flow as

$$Q_{max} = C_v X_{max} \sqrt{P_s} \qquad (7.8)$$

If we now divide (7.6) by (7.8), we obtain a more general, dimensionless valve flow equation

$$\left(\frac{Q}{Q_{max}}\right) = \left(\frac{X}{X_{max}}\right)\sqrt{1 - \left(\frac{P_1}{P_s}\right)} \qquad (7.9)$$

Figure 7.5 shows a plot of equation (7.9). It shows the dimensionless flow through a single port of a hydraulic valve as a function of the dimensionless downstream pressure and the dimensionless opening of the valve.

In a manner similar to that used above, we can write the flow equation for Port B of the valve as

$$Q = C_d W X \sqrt{\frac{2}{\rho}(P_2 - P_d)}$$

If we assume the drain pressure is zero and use the valve constant, we can rewrite this equation as

$$Q = C_v X \sqrt{P_2} \qquad (7.10)$$

Notice that by using the same symbol for flow in this equation as I used in (7.6), I have assumed that the fluid is incompressible and that there is no leakage between one chamber of the ram and the other. That is, the fluid flowing out of the valve at Port A and into the ram equals the fluid flowing out of the other side of the ram and back into the valve at Port B.

Many manufacturers of hydraulic valves list the characteristics of their valves in terms of a load pressure, P_L. The load pressure is defined as

$$P_L = P_1 - P_2 \qquad (7.11)$$

We can solve equations (7.6) and (7.10) for P_1 and P_2, respectively and substitute them into (7.11) as follows

$$P_1 = P_s - \left(\frac{Q}{C_v X}\right)^2 \qquad \text{(7.6) solved for } P_1$$

$$P_2 = \left(\frac{Q}{C_v X}\right)^2 \qquad \text{(7.10) solved for } P_2$$

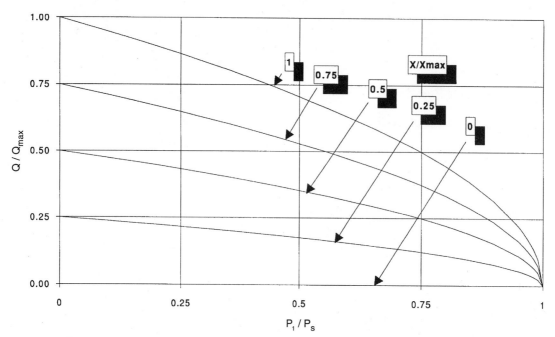

Figure 7.5. Dimensionless flow versus pressure and opening for a single-valve orifice.

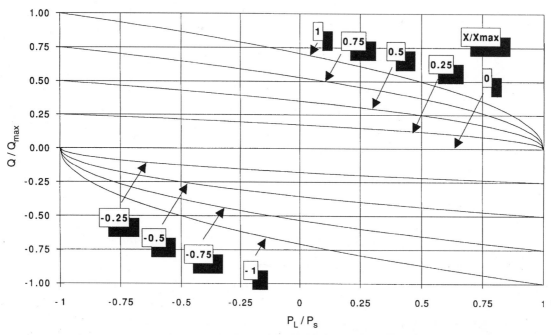

Figure 7.6. Dimensionless flow versus pressure and opening for a four-way hydraulic valve.

Substitute into (7.11)

$$P_L = P_s - \left(\frac{Q}{C_v X}\right)^2 - \left(\frac{Q}{C_v X}\right)^2 = P_s - 2\left(\frac{Q}{C_v X}\right)^2$$

which can then be rearranged to give

$$Q = C_v X \sqrt{\frac{1}{2}(P_s - P_L)} \qquad (7.12)$$

If we divide equation (7.12) by (7.8), we arrive at a dimensionless equation for the flow of a four-way valve

$$\left(\frac{Q}{Q_{max}}\right) = \left(\frac{X}{X_{max}}\right)\sqrt{\frac{1}{2}\left[1 - \left(\frac{P_L}{P_s}\right)\right]}$$

$$(7.13)$$

Figure 7.6 shows this equation in graphical form. Note that this figure includes negative as well as positive valve displacements. Note also that the load pressure could reach a negative value equal to the supply pressure.

Steady-state Model for the Hydraulic Actuator

Now let's develop a math model for the (hydraulic actuator) ram. If the fluid is indeed incompressible and if there is no leakage across the piston, then the product of the velocity of the piston and its cross-sectional area must equal the fluid flow into the ram. That is,

$$Q = A_p \dot{Y} \qquad (7.14)$$

Note that the area of the piston referred to in (7.14) is the cross-sectional area associated with the outer diameter of the piston *less* the cross-sectional area of the piston rod.

We can write the following equation from a free-body diagram of the piston:

$$F_L = P_1 A_p - P_2 A_p = A_p(P_1 - P_2) \quad (7.15)$$

where F_L = load force. Once again note that the area of the piston referred to in (7.15) is the cross-sectional area associated with the outer diameter of the piston *less* the cross-sectional area of the piston rod. We can also write (7.15) in the form

$$F_L = A_p P_L \qquad (7.16)$$

where $P_L = P_1 - P_2$ as defined previously.

Take a close look at (7.14) and (7.16). Q and P_L are the through and across variables of a fluid system and F_L and $\dot{Y}$ are the through and across variables of a mechanical (translational) system. Consequently, these two equations are a math model for a fluid–mechanical transmission or linear hydraulic motor. A hydraulic ram takes in fluid power and converts it to translational mechanical power. As you might imagine, there are also rotary hydraulic motors. They convert fluid power to rotary mechanical power. The source of the hydraulic power is the hydraulic power supply. Thinking of the actuator as a motor will allow you to make easy comparisons with the case study work we did in Chapter 5 where we used DC permanent magnet motors.

Steady-state Model for the Combined Hydraulic Valve and Actuator

Equation (7.12), representing our math model of the valve, and equations (7.14) and (7.16), representing our math model of the actuator, can now be combined as follows:

$$Q = C_v X \sqrt{\frac{1}{2}(P_s - P_L)}$$

$$A_p \dot{Y} = C_v X \sqrt{\frac{1}{2}\left(P_s - \frac{F_L}{A_p}\right)}$$

$$\dot{Y} = \frac{C_v X}{A_p} \sqrt{\frac{1}{2}\left(P_s - \frac{F_L}{A_p}\right)} \qquad (7.17)$$

This math model completely describes the steady-state velocity of the ram as a function of the valve opening, the load, and a number of particulars related to the valve, the ram, and the hydraulic supply pressure. Since we were given the peak load by the ocean engineers and we can compute the peak velocity from the stroke data, then we could use (7.17) to select the supply pressure and the size of the valve and actuator. We will do this shortly, but before we do, let's also make equation (7.17) dimensionless as we did for the valve in (7.13).

We can define the maximum piston velocity using (7.14)

$$Q_{max} = A_p \dot{Y}_{max} \qquad (7.18)$$

and the maximum load force using (7.16)

$$F_{L\,max} = A_p P_s \qquad (7.19)$$

Equations (7.14), (7.16), (7.18) and (7.19) can then be substituted into (7.13) to give

$$\left(\frac{\dot{Y}}{\dot{Y}_{max}}\right) = \left(\frac{X}{X_{max}}\right) \sqrt{\frac{1}{2}\left[1 - \left(\frac{F_L}{F_{L\,max}}\right)\right]} \qquad (7.20)$$

Compare (7.20) with (7.13). You can see that they are identical except for the change of variables. The transmission equations given by (7.14) and (7.16) simply act as scaling factors, allowing us to transform the axis of the valve curves given in Figure 7.6 into the velocity and force variables associated with the translational mechanical system. This is precisely what we did in Chapter 5 when we used the transmission equations of the pulley to allow us to superimpose the shaker table loads onto the torque-speed characteristics of the DC motor.

Sizing the Hydraulic Valve and Actuator

You will recall in Chapter 5, when we sized the motor used with the shaker table, I indicated it was important from an efficiency point of view to get the steady-state load to correspond to the maximum efficiency point of the motor. We can find this point for our linear hydraulic motor using equation (7.20) and the equation for the translational mechanical power

$$Pow = F_L \dot{Y} \qquad (7.21)$$

Substituting (7.20) into this equation gives

$$Pow = F_L \dot{Y}_{max} \frac{X}{X_{max}} \sqrt{\frac{1}{2}\left(1 - \frac{F_L}{F_{L\,max}}\right)} \qquad (7.22)$$

This equation can be arranged in the dimensionless form

$$\left(\frac{Pow}{F_{L\max}\dot{Y}_{\max}}\right)=\left(\frac{F_L}{F_{L\max}}\right)\left(\frac{X}{X_{\max}}\right)\sqrt{\frac{1}{2}\left[1-\left(\frac{F_L}{F_{L\max}}\right)\right]}$$

(7.23)

You can see that (7.23) could produce a family of power curves with each curve corresponding to a particular valve opening. We are interested only in the one corresponding to the maximum opening of the valve. That is where

$$\left(\frac{X}{X_{\max}}\right)=1$$

(7.24)

Substituting (7.24) into (7.23) gives

$$\left(\frac{Pow}{F_{L\max}\dot{Y}_{\max}}\right)=\left(\frac{F_L}{F_{L\max}}\right)\sqrt{\frac{1}{2}\left[1-\left(\frac{F_L}{F_{L\max}}\right)\right]}$$

(7.25)

The condition for maximum power can be obtained by plotting (7.25) or by differentiating (7.25) with respect to $(F_L/F_{L\max})$, setting the result to zero and then solving for $(F_L/F_{L\max})$. Figure 7.7 shows a plot of the equation. You can see that the maximum power point occurs when $(F_L/F_{L\max})$ is around 0.7 and the dimensionless peak power is about 0.27. If you carry out the differentiation, you will find the exact value is

$$\frac{F_L}{F_{L\max}}=\frac{2}{3}$$

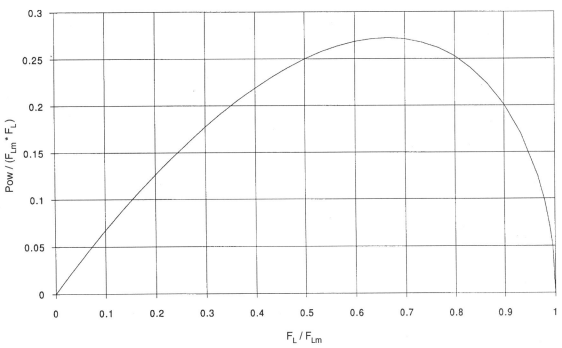

Figure 7.7. Power versus load for a four-way valve and actuator.

and the exact maximum power is

$$Pow = \frac{2}{3}\sqrt{\frac{1}{6}}F_{L\max}\dot{Y}_{\max} = 0.2722 F_{L\max}\dot{Y}_{\max}$$

So, we now need to determine $F_{L\max}$ and $\dot{Y}_{\max}$ so the peak loads given to us by the ocean engineers results in $(F_L / F_{L\max})$ being nearly equal to 2/3. Substituting $(F_L / F_{L\max}) = 2/3$ into (7.20) gives

$$\frac{\dot{Y}}{\dot{Y}_{\max}} = \sqrt{\frac{1}{2}\left(1 - \frac{F_L}{F_{L\max}}\right)}$$

$$= \sqrt{\frac{1}{2}\left(1 - \frac{2}{3}\right)} = 0.4082$$

or

$$\dot{Y}_{\max} = \frac{\dot{Y}}{0.4082}$$

and from (7.18)

$$\dot{Y}_{\max} = \frac{Q_{\max}}{A_P} = \frac{\dot{Y}}{0.4082} \qquad (7.26)$$

The peak load velocity of the flap, $\dot{Y}$, can be obtained by differentiating the stroke equation

$$S = S_o \sin\left(\frac{2\pi}{T}t\right)$$

That is

$$\dot{Y} = \left|\frac{dS}{dt}\right| = \left|\frac{2\pi}{T}S_o \cos\left(\frac{2\pi}{T}t\right)\right| = \frac{2\pi}{T}S_o$$

Substituting values from the table of load data gives

$$\dot{Y} = \frac{2\pi}{T}S_o = \frac{2 \times 3.14}{2.4} \times 25.984$$

$$= 68.03 \text{ in / sec}$$

The maximum possible ram velocity, obtained by substituting into (7.26), is

$$\frac{Q_{\max}}{A_P} = \frac{\dot{Y}}{0.4082} = \frac{68.03}{0.4082}$$

$$= 166.64 \text{ in / sec}$$

or

$$Q_{\max} = 166.64 A_P \qquad (7.27)$$

Since we selected

$$\left(\frac{F_L}{F_{L\max}}\right) = \frac{2}{3}$$

then the maximum force is equal to

$$F_{L\max} = \frac{F_L}{2/3}$$

We can obtain F_L from the data provided by the ocean engineers. Since F_R and F_I are 90 degrees out of phase, the amplitude of the combined load is given by

$$F_L = \sqrt{(F_R)^2 + (F_I)^2}$$

or

$$F_L = \sqrt{(616)^2 + (15.7)^2} = 616.2$$

The maximum force is then equal to

$$F_{L\max} = \frac{F_L}{2/3} = \frac{616}{2/3} = 924 \text{ lbs} = P_S A_P$$

or

$$A_P = \frac{924}{P_S} \qquad (7.28)$$

Equations (7.27) and (7.28) can now be used to select a valve, a ram, and a supply pressure for the hydraulic fluid. When selecting the ram, consideration must be given to the diameter of the piston rod so it will not buckle or excessively bend under load. We'll presume that such calculations have been conducted and that we must have a one-inch diameter piston rod. Since the area of the piston is

$$A_P = \frac{\pi}{4}\left(D^2 - d^2\right)$$

where D is the diameter of the piston and d is the diameter of the piston rod, then the required diameter of the piston is

$$D = \sqrt{\frac{4A_P}{\pi} + d^2} = \sqrt{1.273A_P - 1} \quad (7.29)$$

We can now use (7.27), (7.28) and (7.29) to create a table of required piston diameters and flow rates at several possible supply pressures, as shown in Table 7.2.

Table 7.2.
Piston Diameters and Flow Rates.

P_S (psi)	A_P (sq. in)	Q_{max} (cu. in /sec)	D (in)
3000	0.308	51.33	1.18
2500	0.370	61.66	1.21
2000	0.462	76.99	1.26
1500	0.616	102.65	1.34

At this point we would look into catalogs of hydraulic actuator manufacturers. I did this, and found an actuator with a piston diameter of 1.25

inches and a piston rod of 1-inch diameter. This is close to the 1.26-diameter piston in our table. The catalog also indicated that the actuator could be safely used with a 2000-psi hydraulic supply pressure. We tentatively select this actuator and note the piston area of ram is 0.4418 sq. in.

I also searched through catalogs of hydraulic valve manufacturers and found one manufacturer who supplies electrically actuated four-way hydraulic valves. This manufacturer provides the following formula for computing the flow of the valves

$$Q_R = KI_m\sqrt{P_v} \quad (7.30)$$

where $\quad Q_R$ = rated flow [gallon per minute (gpm)]

$\quad\quad K$ = a valve constant

$\quad\quad I_m$ = the maximum DC current to the valve (15 milliamps)

$\quad\quad P_v = P_s - P_L$ = the valve pressure drop (psi)

The manufacturer rates his valves assuming a 1000-psi drop across the valve; that is, $P_L = 1000$ psi. He has valves rated at 1, 2.5, 5, 10, and 15 gpm. We now need to determine if any of these valves will work in our design. To do this, rearrange (7.30) and substitute values for the rated flow of each valve. Starting with the 10-gpm (38.5 cu.in/sec) valve

$$K = \frac{Q_R}{I_m\sqrt{P_v}} = \frac{38.5}{15\sqrt{1000}} = \frac{1.2175}{15}$$

$$= 0.0812 \frac{\text{cu.in / sec}}{\text{mA - psi}^{1/2}}$$

For our application, the peak load pressure will be

$$P_L = \frac{F_L}{A_p} = \frac{616.2}{0.4418} = 1394.75 \text{ psi}$$

To determine if the valve can supply the desired flow, we substitute the known values into (7.30) and obtain the flow through the valve

$$Q = K I_m \sqrt{P_v}$$
$$= 1.2175 \sqrt{2000 - 1394.75}$$
$$= 29.95$$

The required flow rate is

$$Q = \dot{Y} A_p = 68.03 \times 0.4418$$
$$= 30.06 \text{ cu. in / sec}$$

The required flow is just slightly more than the flow through the valve, but it is close enough. If you investigate the other valves, you will find all those with a rated flow less than 10 gpm will not work at a supply pressure of 2000 psi. The 15-gpm valve will work, but it costs more than the 10-gpm valve and has slower response characteristics. We conclude that the 10-gpm valve is the right choice.

Figure 7.8 shows a summary of our design work. The approximate maximum load path is shown along with the maximum valve-actuator linear hydraulic motor characteristics. The design point we just developed is marked on this figure. In operation, the current to the valve will be adjusted so the motor characteristics continually intersect with the load path.

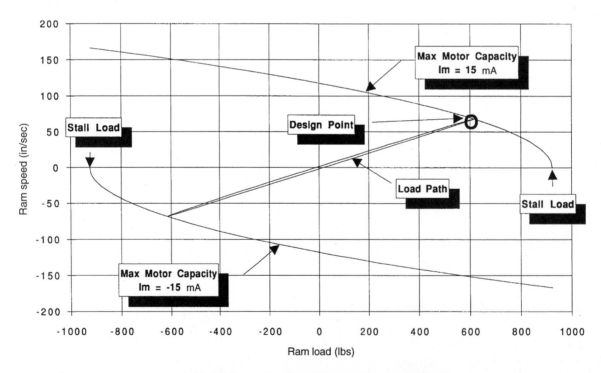

Figure 7.8. Hydraulic motor and load.

Derivation of Dynamic Math Model of System

So far we have only investigated the steady-state behavior of the system at its maximum loading point. In general, this is always the first step in a design process and allows equipment to be approximately sized. The next step is to investigate the dynamic performance of the system to ensure the steady-state design is adequate. In this case study we have also been asked to provide the ocean engineers with the transfer function between the input voltage to the valve and the output stroke of the wave flap.

Dynamic Model for the Hydraulic Valve

As you saw previously, equation (7.30) is a steady-state equation for the maximum flow through an electrically actuated hydraulic valve. It can be rewritten for any flow using

$$Q = KI\sqrt{P_v} = KI\sqrt{P_S - P_L} \qquad (7.31)$$

where

$$K = 0.0812 = \frac{\text{cu.in / sec}}{\text{mA - psi}^{1/2}}$$

as we determined earlier.

Electrically actuated hydraulic valves generally contain a small torque motor that drives a small pilot valve, which in turn positions the main spool valve. Manufacturers will supply *step response* and or *frequency response* data for their valves, expecting you as a designer to know what they mean and how to use them. Some typical examples of manufacturer's dynamic data for hydraulic valves are shown in Figure 7.9.

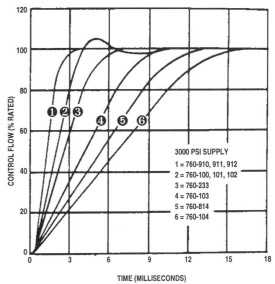

Typical transient response of 760 Servovalves is given in the figure above. The straight-line portion of the response represents saturation flow from the pilot state which will vary with the square root of the change in supply pressure.

(a). *Step Responses*

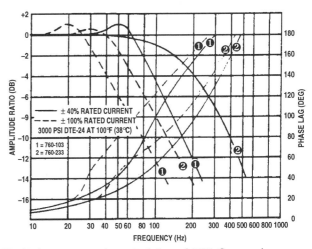

Typical response characteristics of 760 Servovalves are given in the figure above. Servovalve frequency response will vary with signal amplitude, supply pressure, temperature, and internal valve design parameters.

(b). *Frequency Responses*

Figure 7.9. Typical examples of manufacturers' dynamic data for hydraulic valves.

We can make use of this data to help us build an approximate dynamic model of the valve. The torque motor consists of an electromagnet. That is, it is a coil that contains an and a resistance in series as shown in Figure 7.10. By now you should be able to write the equation relating the input voltage to the current of this circuit by inspection. The sum of the voltages around the loop must equal zero, so

$$E - L\frac{dI}{dt} - RI = 0$$

which can be rearranged to

$$L\frac{dI}{dt} + RI = E \qquad (7.32)$$

You will immediately recognize this equation as a first-order, linear, ordinary differential equation with constant coefficients. It can be rewritten as

$$\tau\frac{dI}{dt} + I = \frac{1}{R}E$$

or

$$(\tau D + 1)I = \frac{1}{R}E \qquad (7.33)$$

where $\tau = \dfrac{L}{R}$ is the time constant of the circuit.

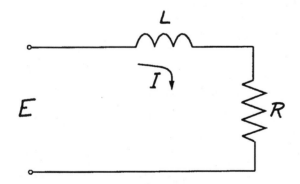

Figure 7.10. Schematic model of torque motor in hydraulic valve.

We can use the manufacturer's data and our knowledge of the response of a first-order system to determine the time constant so our math model best fits the manufacturer's data. From the step response graph we pick off the following:

Percent Rated Flow	Time (seconds)	Estimated time constant (seconds)
63.2	0.005 (1 τ)	0.005
86.5	0.007 (2 τ)	0.0035
95.0	0.009 (3 τ)	0.003
		0.0038 (avg)

We can also estimate the time constant from the break frequency of the frequency response data, by using the averages of the two quoted ratings, 100% and 40%. Using the ±100% rated current, the break frequency is about 20 Hz. Therefore,

$$\tau = \frac{1}{2\pi f} = \frac{1}{2\pi(20)} = 0.008 \text{ secs}$$

Using the ±40% rated current data, the break frequency is around 50 Hz, giving an estimate for the time constant of 0.003 seconds. These two break frequency estimates give an average of .0055 seconds. Averaging the step response and the break frequency estimates gives 0.0045 seconds for the time constant, which we round to 0.005 seconds.

We also note from the manufacturer's data that the resistance of the coils is 800 ohms. We can then write equation (7.33) as

$$0.005\frac{dI}{dt} + I = \frac{1}{800}E \qquad (7.34)$$

Note that in steady-state ($dI/dt = 0$) equation (7.34) indicates that ±12 volts DC must be input to the valve to give ±15 mA of current. That is,

$$E = 800 \times \pm 0.015 = \pm 12 \text{ volts}$$

Equations (7.31) and (7.34) make up a dynamic math model of the valve. Equation (7.34) provides the valve current as a function of the input voltage and equation (7.31) provides the output flow given the current, the supply pressure, and the load pressure. We can show these two equations in block diagram notation as in Figure 7.11. Note that I made up a nonlinear block (square root) to show the nonlinear nature of equation (7.33).

Dynamic Model for the Hydraulic Actuator

Equations (7.14) and (7.16) comprise the math model of the actuator. We can use (7.14) to obtain dY/dt given Q and (7.16) to determine P_L given F_L. The actuator model has been added to the block diagram in Figure 7.12.

Note in this figure that I have shown a box labeled *Load* with dY/dt as the input and F_L as the output. We really don't know what this block looks like mathematically because the ocean engineers only gave us the results of their calculations in tabular form.

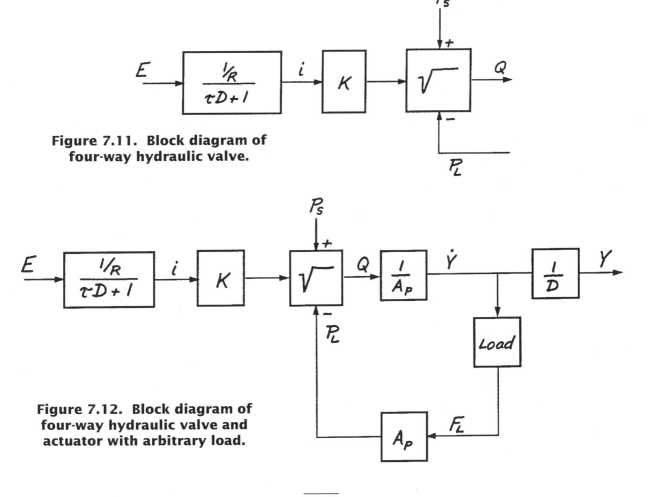

Figure 7.11. Block diagram of four-way hydraulic valve.

Figure 7.12. Block diagram of four-way hydraulic valve and actuator with arbitrary load.

Linearization for Obtaining a Linear Dynamic Model

The block diagram shown in Figure 7.12 is not what you are used to seeing. It has a nonlinear square-root block and a block describing loads given to us in a tabular form. To gain a clearer understanding of the system dynamics, it is often helpful to linearize functions about steady-state operating points. The resulting linear equations are then applicable for small excursions about the steady-state operating point. This is an extremely useful concept, so I will go through the process step by step.

We'll start by building a linearized model of the hydraulic valve. The math model we've developed so far is comprised of two equations:

$$\tau \frac{dI}{dt} + I = \frac{1}{R} E \qquad (7.33) \text{ repeated}$$

and

$$Q = KI\sqrt{P_S - P_L} = f(I, P_L) \qquad (7.31) \text{ repeated}$$

Equation (7.33) describes the valve torque motor current as a function of the input voltage and equation (7.31) describes the flow as a function of the current and load pressure. Equation (7.33) is linear, equation (7.31) is not.

I've written (7.31) in functional notation to indicate that it has two variables, I and P_L, and two constants, K and P_S. The equation is nonlinear because it involves *taking the square root of* P_L and involves the *multiplication of two variables*, I and the square root of $(P_S - P_L)$. Equation (7.31) can be made linear by expanding the

nonlinear function $Q = f(I, P_L)$ about a steady-state operating point (I_0, P_{L0}) using the Taylor's Series (discussed in Appendix A, if you need a brush-up). First write the expansion in functional notational form

$$Q = f(I_0, P_{L0}) + \left.\frac{\partial f}{\partial I}\right|_{I_0, P_{L0}} (I - I_0)$$

$$+ \left.\frac{\partial f}{\partial P_L}\right|_{I_0, P_{L0}} (P_L - P_{L0}) \qquad (7.35)$$

where

I_0 = the steady-state value of the current

P_{L0} = the steady-state value of the load pressure

$f(I_0, P_{L0})$ = the function given by (7.31) evaluated at I_0 and P_{L0}

$\left.\frac{\partial f}{\partial I}\right|_{I_0, P_{L0}}$ = the partial derivative of (7.31) with respect to I evaluated at I_0 and P_{L0}

$\left.\frac{\partial f}{\partial P_L}\right|_{I_0, P_{L0}}$ = the partial derivative of (7.31) with respect to P_L evaluated at I_0 and P_{L0}

$(I - I_0)$ = the deviation of the current from its steady-state operating point

$(P_L - P_{L0})$ = the deviation of the load pressure from its steady-state operating point.

Equation (7.35) looks a lot more complicated than it really is. It can be simplified by defining

$$f(I_0, P_{L0}) = Q_0$$

$$(I - I_0) = i$$

$$(P_L - P_{L0}) = p_L$$

$$(Q - Q_0) = q$$

Now we can rewrite (7.35) as

$$q = \left.\frac{\partial f}{\partial I}\right|_{I_0, P_{L0}} i + \left.\frac{\partial f}{\partial P_L}\right|_{I_0, P_{L0}} p_L$$

$$= C_1 i + C_2 p_L \qquad (7.36)$$

where the constants are given by

$$C_1 = \left.\frac{\partial f}{\partial I}\right|_{I_0, P_{L0}}$$

and

$$C_2 = \left.\frac{\partial f}{\partial P_L}\right|_{I_0, P_{L0}}$$

Equation (7.36) is now a linear equation. The two constants, C_1 and C_2, are obtained by taking the partial derivatives and evaluating these at the steady-state operating point.

Taking partial derivatives is no big deal. Simply treat one variable as a constant and then take the derivative of the function with respect to the other. Then repeat with the other variable. That is,

$$\frac{\partial f}{\partial I} = \frac{\partial \left(KI\sqrt{P_S - P_L}\right)}{\partial I} = \left(K\sqrt{P_S - P_L}\right)\frac{dI}{dt}$$

$$= K\sqrt{P_S - P_L}$$

and

$$\frac{\partial f}{\partial P_L} = \frac{\partial \left(KI\sqrt{P_S - P_L}\right)}{\partial P_L} = (KI)\frac{d\left(P_S - P_L\right)^{1/2}}{dP_L}$$

$$= (KI)\frac{1}{2}\left(P_S - P_L\right)^{-1/2}\frac{d\left(P_S - P_L\right)}{dP_L}$$

$$= (KI)\frac{1}{2}\left(P_S - P_L\right)^{-1/2}(-1)$$

Now evaluate these partial derivatives at the operating point and get

$$C_1 = \left.\frac{\partial f}{\partial I}\right|_{I_0, P_{L0}} = K\sqrt{P_S - P_{L0}} \qquad (7.37)$$

$$C_2 = \left.\frac{\partial f}{\partial P_L}\right|_{I_0, P_{L0}} = (KI_0)\frac{1}{2}\left(P_S - P_{L0}\right)^{-1/2}(-1)$$

$$= -\frac{KI_0}{2\sqrt{P_S - P_{L0}}} \qquad (7.38)$$

Take a moment and look at these constants. Notice that C_1 does not involve I_0 and is always positive regardless of the value of P_{L0}. Notice that C_2 involves both I_0 and P_{L0} and is always negative. Since C_2 is always negative, we'll rewrite this coefficient as

$$b_2 = -C_2 = \frac{KI_0}{2\sqrt{P_S - P_{L0}}} \qquad (7.39)$$

Our linearized version of equation (7.31) then becomes

$$\boxed{q = C_1 i - b_2 p_L} \qquad (7.40)$$

where C_1 is given by (7.38) and b_2 is given by (7.39)

We can cast equation (7.33) in the form of excursions about the steady-state operating point also. First we introduce the variables

$$i = I - I_0 \quad \Rightarrow \quad I = i + I_0 \qquad (7.41)$$

and

$$e = E - E_0 \quad \Rightarrow \quad E = e + E_0 \qquad (7.42)$$

Now substitute (7.41) and (7.42) into (7.33)

$$\tau\left(\frac{di}{dt} + \frac{dI_0}{dt}\right) + i + I_0 = \frac{1}{R}e + \frac{1}{R}E_0 \qquad (7.43)$$

In steady-state,

$$\frac{dI_0}{dt} = 0$$

and

$$I_0 = \frac{1}{R}E_0$$

Substituting these into (7.43) results in

$$\boxed{\tau\frac{di}{dt} + i = \frac{1}{R}e} \qquad (7.44)$$

Equation (7.40) and (7.44) now constitute a linearized model of the electrically actuated hydraulic valve. A block diagram of this linearized valve is shown in Figure 7.13.

Equations (7.14) and (7.16) are our math model for the actuator.

$$Q = A_p \dot{Y} \qquad \text{(7.14) repeated}$$

$$F_L = A_p P_L \qquad \text{(7.16) repeated}$$

The equations are already linear, so it's easy to write the equations in terms of excursions about the steady-state operating points. Rewrite (7.14) as

$$\dot{Y} = \frac{Q}{A_p}$$

then

$$\dot{Y}_0 = \frac{Q_0}{A_p}$$

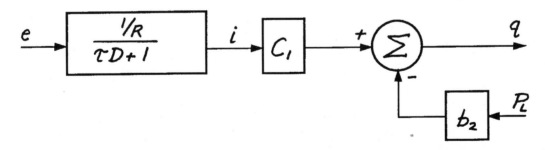

Figure 7.13. Linearized electrically actuated four-way hydraulic valve.

so we can write

$$\dot{y} = \dot{Y} - \dot{Y}_0 = \frac{Q}{A_P} - \frac{Q_0}{A_P}$$

or

$$\dot{y} = \frac{1}{A_P} q \qquad (7.45)$$

In a similar manner we have from (7.14)

$$f_L = F_L - F_{L0} = A_P P_L - A_P P_{L0} = A_P p_L$$

or

$$p_L = \frac{1}{A_P} f_L \qquad (7.46)$$

Figure 7.14 shows (7.45) and (7.45) added to the block diagram. The figure now shows a complete block diagram of a linearized hydraulic valve and actuator. All that is missing is a relationship for the load.

The load on an actuator typically depends on the nature of the mechanical load. The most general linear load is a combination of a mass, a damper, and a spring. That is,

$$F_L = m\ddot{Y} + b\dot{Y} + kY \qquad (7.47)$$

In our wave flap we have no spring loads so (7.47) reduces to

$$F_L = m\ddot{Y} + b\dot{Y} \qquad (7.48)$$

We can cast this equation in the form of excursions about an operating point by introducing the variables

$$\dot{y} = \dot{Y} - \dot{Y}_0 \Rightarrow \dot{Y} = \dot{y} + \dot{Y}_0 \Rightarrow \ddot{Y} = \ddot{y} + \ddot{Y}_0 \quad (7.49)$$

$$f_L = F_L - F_{L0} \Rightarrow F_L = f_L + F_{L0} \quad (7.50)$$

Substituting these into (7.48) and simplifying gives

$$f_L + F_{L0} = m\left(\ddot{y} + \ddot{Y}_0\right) + b\left(\dot{y} + \dot{Y}_0\right)$$
$$= m\ddot{y} + b\ddot{y} + \left(m\ddot{Y}_0 + \dot{Y}_0\right)$$

or

$$f_L = m\ddot{y} + b\dot{y} \qquad (7.51)$$

since

$$F_{L0} = m\ddot{Y}_0 + \dot{Y}_0$$

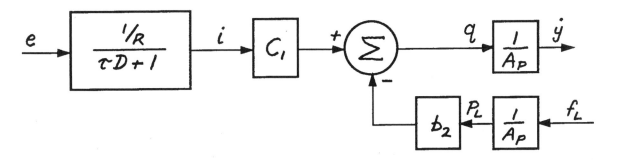

Figure 7.14. Linearized electrically actuated four-way hydraulic valve and double-acting hydraulic actuator.

Case $m = 0$

Let's first look at the case where m is zero. This case is similar to our design load case because F_R is so much larger than F_I. Equation (7.51) then reduces to

$$f_L = b\dot{y} \qquad (7.52)$$

The load force is directly proportional to the velocity. We can add (7.52) to our block diagram as shown in Figure 7.15.

See how nicely this all fits together? You can clearly see the relationship between the ram velocity and the input voltage. You can also see how the fluid *across* and *through* variables (q and p_L) are transformed into the mechanical *across* and *through* variables (dy / dt and f_L) by the actuator.

The block diagram can be reduced by combining equations (7.40), (7.44), (7.45), (7.46) and (7.52) to obtain the transfer function between e and dy / dt. Combining equations (7.40), (7.45), (7.46), and (7.52) gives

$$\dot{y} = \frac{C_1}{\left(\dfrac{A_P^2 + b_2 b}{A_P}\right)} i$$

or

$$\dot{y} = C_3 i \qquad (7.53)$$

where

$$C_3 = \frac{C_1}{\left(\dfrac{A_P^2 + b_2 b}{A_P}\right)} \qquad (7.54)$$

which shows us that the velocity of the ram is directly proportional to the current flowing into

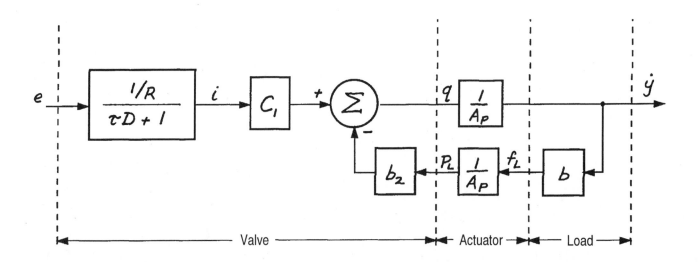

Figure 7.15. Linearized electrically actuated four-way hydraulic valve, double-acting hydraulic actuator, and a velocity-dependent load.

the valve. Using (7.53) allows us to reduce the block diagram of Figure 7.15 to that shown in Figure 7.16. You will immediately recognize this transfer function as a first-order, linear, ordinary differential equation with constant coefficients:

$$\tau \frac{dv}{dt} + v = \frac{C_3}{R} e \qquad (7.55)$$

Notice that for the case we are now looking at, where the load is proportional to velocity, the

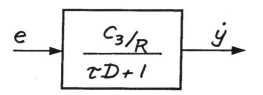

Figure 7.16. Transfer function for the block diagram shown in Figure 7.15.

time constant of the system is determined solely by the valve.

We are primarily interested in the transfer function between the input voltage and the output displacement of the ram, not its velocity. In operational notation

$$y = \frac{1}{D} \dot{y} \qquad (7.56)$$

We can add this block to our block diagram as shown in Figure 7.17 and derive the new transfer function shown. This results in a second-order, linear, ordinary differential equation with constant coefficients.

$$\tau \frac{d^2 y}{dt^2} + \frac{dy}{dt} = \frac{C_3}{R} e \qquad (7.57)$$

Note that the solution of this equation is easily obtained. You first obtain the solution to (7.55) and then integrate the result.

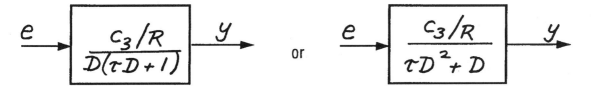

Figure 7.17. Transfer function between input voltage and output displacement for a linearized electrically actuated four-way hydraulic valve, double-acting hydraulic actuator, and a velocity-dependent load.

Case $m \neq 0$

Let's now return to the case where the mass of the paddle and the hydrodynamic forces proportional to flap acceleration are included in the load. That is, when

$$f_L = m\ddot{y} + b\dot{y} \qquad (7.58)$$

If we now combine equations (7.40), (7.45), (7.46), and (7.58) we obtain

$$\left(\frac{b_2 m}{A_P}\right)\ddot{y} + \left(\frac{A_P^2 + b_2 b}{A_P}\right)\dot{y} = C_1 i$$

or

$$C_4\ddot{y} + \dot{y} = C_3 i \implies (C_4 D + 1)\dot{y} = C_3 i \quad (7.59)$$

where

$$C_4 = \frac{b_2 m}{A_P^2 + b_2 b} \qquad (7.60)$$

and where C_3 was given previously by (7.54).

Figure 7.18 shows the block diagram of our system when the load has both velocity- and acceleration-dependent terms. Compare this with the transfer function given in Figure 7.16. Note that if C_4 is zero ($m = 0$) then the block diagram in Figure 7.18 reduces to the one in Figure 7.16. You should also note that the addition of a load term proportional to acceleration has made the equation relating the input voltage to the output velocity second-order. That is,

from Figure 7.18 we can see that the velocity ($v = dy / dt$) of the ram is described by the following second-order, linear, ordinary differential equation with constant coefficients:

$$C_4\tau\frac{d^2 v}{dt^2} + (C_4 + \tau)\frac{dv}{dt} + v = \frac{C_3}{R}e \qquad (7.61)$$

where C_3 is given by (7.54), C_4 by (7.60) and τ by (7.33).

The transfer function between the input voltage and the output displacement of the ram is shown in Figure 7.19. This transfer function results in a third-order, linear, ordinary differential equation with constant coefficients.

$$C_4\tau\frac{d^3 v}{dt^3} + (C_4 + \tau)\frac{d^2 y}{dt^2} + \frac{dy}{dt} = \frac{C_3}{R}e \qquad (7.62)$$

This equation can also be solved by first finding the solution to (7.61) and then integrating the result.

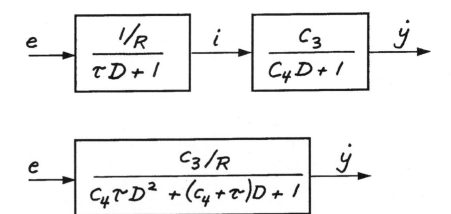

Figure 7.18. Block diagram of system when load has both velocity- and acceleration-dependent terms.

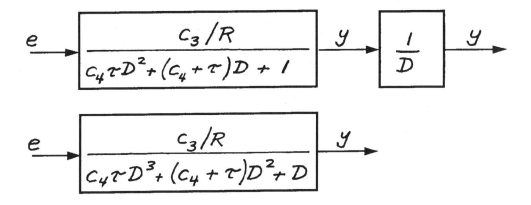

Figure 7.19. Transfer function between input voltage and output displacement for a linearized electrically actuated four-way hydraulic valve, double-acting hydraulic actuator, and a load proportional to velocity and acceleration.

Introduction to Feedback Control Systems

In the development of the math model for this system, you discovered that the velocity of the ram is proportional to the current supplied to the valve. If we implemented this system, we would find the ram would drift. That's because any small error in the zero voltage generated by the computer would produce a small ram velocity, causing the ram to move toward one of its extreme positions. This is obviously not what the ocean engineers wanted.

The problem can easily be solved by adding a sensor that will provide a voltage proportional to the position of the ram. This voltage can then be compared with the voltage generated by the computer and the difference used to drive the valve. This technique is called *automatic feedback control*. As you will soon see, working with automatic feedback controllers is straightforward.

The feedback concept is shown in Figure 7.20. The sensor used is a linear potentiometer, or *pot*. This device is nothing more than a variable resistor in which the resistance can be varied mechanically with a wiper. We apply a positive voltage to one end of the resistor and an equal but opposite voltage to the other. Then when the wiper is centered, the output signal is zero. The voltage signal from this sensor can be calibrated so it produces a voltage proportional to the position of the ram.

Case *m* = 0

Figure 7.21 shows the block diagram of Figure 7.16 with the feedback scheme implemented. Let's investigate how the feedback affects the system math model. First we write the equation for the voltage applied to the valve e in terms of the computer-generated voltage e_c and the feedback voltage e_f.

$$e = e_c - e_f \qquad (7.63)$$

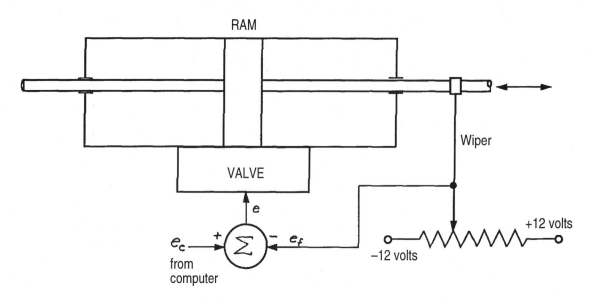

Figure 7.20. Schematic showing automatic feedback control.

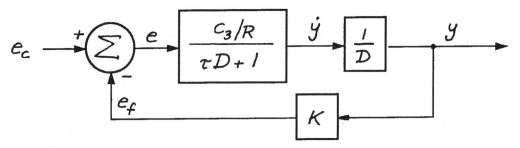

Figure 7.21. Block diagram of Figure 7.17 with feedback control.

Then we write the equations for e_f and e:

$$e_f = Ky \qquad (7.64)$$

and

$$e = \frac{\tau D^2 + D}{C_3/R} y \qquad (7.65)$$

Now we can substitute (7.64) and (7.65) into (7.63) and rearrange terms:

$$\frac{\tau D^2 + D}{C_3/R} y = e_c - Ky$$

$$\left(\frac{\tau D^2 + D}{C_3/R} + K \right) y = e_c$$

$$\left(\tau D^2 + D + \frac{KC_3}{R} \right) y = \frac{C_3}{R} e_c \qquad (7.66)$$

This transfer function gives the following second-order, linear, ordinary differential equation with constant coefficients.

$$\tau \frac{d^2 y}{dt^2} + \frac{dy}{dt} + \frac{KC_3}{R} y = \frac{C_3}{R} e_c \quad (7.67)$$

Compare (7.67) with (7.57). You can see that the equations are identical except now the math model has a term associated with the ram displacement. What we have succeeded in doing with the feedback is to make our system behave as if it had a spring attached to the wave flap. This can be seen when the derivatives are zero as they would be in a steady-state condition. Then (7.67) reduces to

$$\frac{KC_3}{R} y = \frac{C_3}{R} e_c$$

or

$$Ky = e_c \quad (7.68)$$

Equation (7.68) provides us with an equation for setting the value of the feedback gain K.

Since we want the range of y (±25.984 inches for Load Condition 4) to correspond to the range of e_c (±12 volts) we have

$$K = \frac{\Delta e_c}{\Delta y} = \frac{2 \times 12}{2 \times 25.984} = 0.4618 \frac{\text{volts}}{\text{in}} \quad (7.69)$$

Case $m \neq 0$

Figure 7.22 shows the block diagram of Figure 7.18 with the feedback scheme implemented. Let's investigate how the feedback affects that system math model. We proceed exactly as above and write the equation for the voltage applied to the valve e in terms of the computer-generated voltage e_c and the feedback voltage e_f.

$$e = e_c - e_f \quad (7.70)$$

Then we write the equations for e_f and e.

$$e_f = Ky \quad (7.71)$$

and

$$e = \frac{C_4 \tau D^3 + (C_4 + \tau) D^2 + D}{C_3 / R} y \quad (7.72)$$

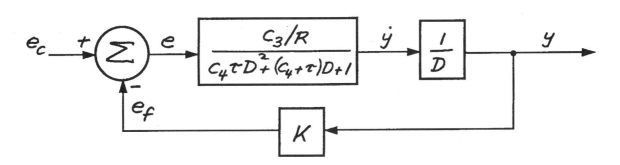

Figure 7.22. Block diagram of Figure 7.19 with feedback control.

Now substitute (7.71) and (7.72) into (7.70) and rearrange terms:

$$\frac{C_4 \tau D^3 + (C_4 + \tau)D^2 + D}{C_3/R}y = e_c - Ky$$

$$\left(\frac{C_4 \tau D^3 + (C_4 + \tau)D^2 + D}{C_3/R} + K\right)y = e_c$$

$$\left(C_4 \tau D^3 + (C_4 + \tau)D^2 + D + \frac{KC_3}{R}\right)y = \frac{C_3}{R}e_c$$

(7.73)

This transfer function gives the following third-order, linear, ordinary differential equation with constant coefficients.

$$C_4 \tau \frac{d^3 y}{dt^3} + (C_4 + \tau)\frac{d^2 y}{dt^2} + \frac{dy}{dt} + \frac{KC_3}{R}y = \frac{C_3}{R}e_c$$

(7.74)

Compare (7.74) with (7.62). You will again see that the equations are identical, except now the math model has a term associated with the ram displacement. We have again succeeded in making our system behave as if it had a spring attached to the wave flap. The steady-state condition again reduces to

$$\frac{KC_3}{R}y = \frac{C_3}{R}e_c$$

or

$$Ky = e_c \qquad (7.75)$$

and the feedback gain K remains the same.

Dynamic Analysis

We are now ready to investigate the dynamic response of our system. Let's begin with our most complicated math model, equation (7.74). We'll first determine numerical values for the coefficients, C_4 and C_3. Values for τ, R, and K have already been found and are equal to 0.005 seconds, 800 ohms (0.800 volts/mA), and 0.4618 volts/in respectively.

The coefficient C_3 is given by (7.54)

$$C_3 = \frac{C_1}{\left(\frac{A_P^2 + b_2 b}{A_P}\right)}$$

where

$$C_1 = K\sqrt{P_S - P_{L0}}$$

$$b_2 = \frac{KI_0}{2\sqrt{P_S - P_{L0}}}$$

$$A_P = 0.4418 \text{ inches}$$

We can determine the coefficient b by dividing the peak resistive load, F_R, by the corresponding peak velocity. For Load Condition 4,

$$b = \frac{F_R}{\left(\frac{2\pi}{T}\right)S_0} = \frac{616}{68.03} = 9.05 \frac{\text{lbs}}{\text{in}/\text{sec}}$$

The coefficient C_4 is given by (7.60) as

$$C_4 = \frac{b_2 m}{A_P^2 + b_2 b}$$

We can determine m as we did b. For Load Condition 4

$$m = \frac{F_I}{\left(\frac{2\pi}{T}\right)^2 S_0} = \frac{15.7}{\left(\frac{2 \times 3.14}{2.4}\right)^2 (25.984)}$$

$$= 0.088 \frac{\text{lbs}}{\text{in / sec}^2}$$

We now need to decide what values to use for the operating points, I_0 and P_{L0}. The wave paddle will be operated with a sinusoidal input voltage. Figure 7.8 shows the load path for this motion. The plot indicates that there really is no steady-state operating point. We could take the origin

$$I_0 = 0$$
$$P_{L0} = 0$$

as the steady-state operating point and treat the sinusoidal motion as excursions about the origin. If we do this then

$$b_2 = 0$$

$$C_4 = 0$$

$$C_1 = K\sqrt{P_s} = 0.0812\sqrt{2000} = 3.63 \ \frac{\text{cu.in / sec}}{\text{mA}}$$

$$C_3 = \frac{C_1}{A_P} = \frac{3.63}{0.4418} = 8.22 \ \frac{\text{cu.in / sec}}{\text{mA - sq.in}} \quad (7.76)$$

With C_4 equal to zero, equation (7.74) reduces to our second-order math model given by (7.67). So we will use

$$\boxed{\tau \frac{d^2 y}{dt^2} + \frac{dy}{dt} + \frac{KC_3}{R} y = \frac{C_3}{R} e_c}$$

$$(7.67)$$
repeated

as the model of our system.

Now let's place (7.67) it in the standard form discussed in Chapter 6. That is,

$$\frac{\tau R}{KC_3} \frac{d^2 y}{dt^2} + \frac{R}{KC_3} \frac{dy}{dt} + y = \frac{1}{K} e_c \quad (7.77)$$

By comparison with the standard form,

$$\frac{1}{\omega_n^2} \frac{d^2 y}{dt^2} + \frac{2\xi}{\omega_n} \frac{dy}{dt} + y = \frac{1}{K} e_c$$

you can see that the natural frequency of the system is

$$\omega_n = \sqrt{\frac{KC_3}{\tau R}} \quad (7.78)$$

and the damping ratio is

$$\xi = \frac{1}{2}\sqrt{\frac{R}{\tau KC_3}} \quad (7.79)$$

Substituting numerical values gives

$$\omega_n = \sqrt{\frac{KC_3}{\tau R}} = \sqrt{\frac{0.4618 \times 8.22}{0.005 \times 0.800}}$$

$$= 30.8 \ \text{rad / sec}$$

and

$$\xi = \frac{1}{2}\sqrt{\frac{R}{\tau KC_3}}$$

$$= \frac{1}{2}\sqrt{\frac{0.800}{0.005 \times 0.4618 \times 8.22}} = 3.25$$

From the value of ξ, we can see that the system is heavily damped. The roots of the characteristic equation are real, not imaginary. They are

$$r_1, r_2 = -\frac{1}{2\tau} \pm \sqrt{\left(\frac{1}{2\tau}\right)^2 - \frac{KC_3}{\tau R}} \qquad (7.90)$$

Substituting numerical values gives

$$r_1, r_2 = -\frac{1}{2 \times 0.005}$$

$$\pm \sqrt{\left(\frac{1}{2 \times 0.005}\right)^2 - \frac{0.4618 \times 8.22}{0.005 \times 0.800}}$$

$$r_1, r_2 = -100 \pm 95.14$$

$$r_1 = -4.86$$

$$r_2 = -195.14$$

From Chapter 6, we can write the response to a step input as

$$y = \frac{1}{K}e_c^* + A_1 e^{-4.86t} + A_2 e^{-195.14t} \qquad (7.91)$$

where e_c^* is the magnitude of the voltage step. The coefficients A_1 and A_2 are determined from the initial conditions $y(t=0)$ and $dy/dt(t=0)$. If we set both of these initial conditions equal to zero and solve for A_1 and A_2, we arrive at

$$y = \frac{e_c^*}{0.4618}\left[1 - 1.0248 e^{-4.86t} + 0.0255 e^{-195.14t}\right] \qquad (7.92)$$

Figure 7.23 shows the response of the wave flap to a step input voltage as given by (7.92). It is important that you notice how this system response resembles the step response of a first-order differential equation. Take notice of the relative contributions of the two roots to the total step response. The root at −195.14 dies away very quickly and makes almost no contribution to the response. The root at −4.86 governs the response and is called the *dominant root*.

Let's investigate the system further to determine what characteristics are controlling these roots. Refer to equation (7.57)

$$\tau\frac{d^2y}{dt^2} + \frac{dy}{dt} = \frac{C_3}{R}e \qquad (7.57)\text{ repeated}$$

This is what the math model looked like before we added the feedback potentiometer. The two roots to this equation are given by (7.90) with $K = 0$. That is

$$r_1, r_2 = -\frac{1}{2\tau} \pm \sqrt{\left(\frac{1}{2\tau}\right)^2} = -\frac{1}{2\tau} \pm \frac{1}{2\tau}$$

$$r_1 = -\frac{1}{\tau} = -\frac{1}{0.005} = -200$$

$$r_2 = 0$$

In Figure 7.24, I have plotted these two roots on the complex plane along with the roots for the system when feedback is present. By considering both sets of roots to be described by (7.90) with K as a parameter, you can see that the presence of the feedback has moved the root at the origin to the left, and the root at −200 to the right. If K were increased further, the roots would continue to move as discussed in Chapter 6.

The fact that the root nearest the origin dominates, or governs, the behavior of the system

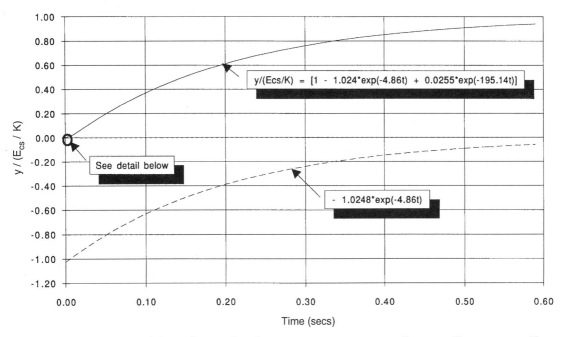

Figure 7.23(a). Dimensionless step response of wave flap controller.

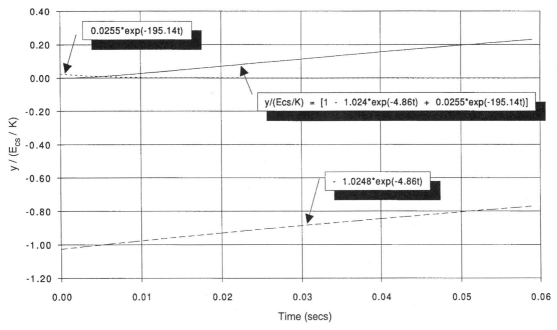

Figure 7.23(b). Detail of dimensionless step response of wave flap controller.

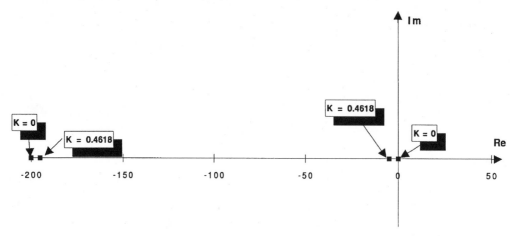

Figure 7.24. Root locus plot showing effect of feedback gain on location of roots.

can be put to good use in simplifying the math model. The time constant of the valve is so fast that it can be considered equal to zero. In other words, the math model of the torque motor given by (7.33) can be simplified to

$$I = \frac{1}{R}E$$

That's equivalent to setting the time constant to zero. Our math model given by (7.67) is therefore reduced to

$$\frac{dy}{dt} + \frac{KC_3}{R}y = \frac{C_3}{R}e_c$$

or

$$\frac{R}{KC_3}\frac{dy}{dt} + y = \frac{1}{K}e_c \qquad (7.93)$$

Equation (7.93) can also be written as

$$\tau\frac{dy}{dt} + y = \frac{1}{K}e_c \qquad (7.94)$$

where

$$\tau = \frac{R}{KC_3} \qquad (7.95)$$

You can see that the time constant of the system employing feedback contains the feedback gain parameter. Numerically, the time constant of this simplified model is

$$\tau = \frac{R}{KC_3} = \frac{0.800}{0.4618 \times 8.22} = 0.2108$$

So the characteristic equation for (7.95) has a single root equal to

$$r = -\frac{1}{\tau} = -\frac{1}{0.2108} = -4.745$$

This is nearly identical to the value of the dominant root ($r_1 = -4.86$) discussed above. The step response solution to (7.93) is

$$y = \frac{e_c^*}{0.4618}\left[1 - e^{-4.745t}\right] \qquad (7.96)$$

If you compare this equation with the solution to the second-order math model given by (7.92), you will see that they are almost identical.

Figure 7.25 shows the comparison of the two solutions. We could give the ocean engineers the transfer function resulting from equation (7.94) to use in their computer codes to correct for discrepancies between the observed wave height and the computer-generated voltage level. That is, we would provide them with

$$\frac{y}{e_c} = \frac{1/K}{\tau D + 1} = \frac{2.1654}{0.2108D + 1} \; \frac{\text{in}}{\text{volt}} \qquad (7.97)$$

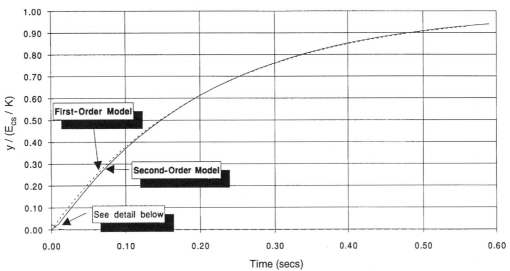

Figure 7.25(a). Comparison of second-order and first-order math models wave flap controller with position feedback.

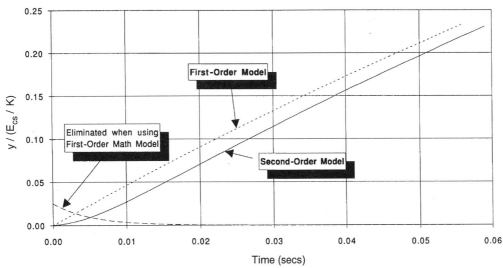

Figure 7.25(b). Detail of comparison of second-order and first-order math models wave flap controller with position feedback.

Chapter

8

Constructing and Analyzing Higher-Order Math Models

Objectives

By the conclusion of this chapter, you will be able to:

- ■ **Develop math models of complex systems.**

- ■ **Solve higher-order math models using numerical techniques.**

- ■ **Analyze higher-order math models using exact solution techniques.**

- ■ **Determine dominant roots of a system math model and learn how to use these to approximate a higher-order math model with a first- or second-order math model.**

- ■ **Determine the exact and approximate frequency response of higher-order math models.**

8.1 Introduction

If you have read the material, studied the examples diligently, and worked the problems that have been presented up to this point, then you now know a great deal about engineering design. By now you know that the first step in engineering design is the development of the simplest possible math model to describe the real engineering system you're trying to design and/or analyze. So far, you've learned how to prepare math models for a wide variety of engineering systems consisting of up to two energy storage elements and one energy dissipative element. You have also learned how to analyze these models. The two case studies given in Chapters 5 and 7 have provided insight into how to apply the modeling and analysis techniques to the solution of practical engineering problems.

You may be wondering at this point if the math modeling and analysis procedures are going to get more and more complicated as the order of the math models increases. The answer is "not really." You have already learned the four basic math model development techniques. Using these tools to formulate math models for systems with numerous energy storage and dissipative elements does tend to get more complicated, but only in regards to the amount of algebra required to derive the math model. Performing lots of algebra tends to lead to mistakes, which means you have to check your work more carefully. But as I have said before, the most difficult part of engineering is creating a math model that properly represents the system you're trying to design and/or analyze. Only through practice will this become easier.

As the math models get more complex, the order of the differential equation increases. Already you have learned that this does not necessarily mean the analysis will be more difficult. For example, you know the numerical solution method allows you to break an *nth*-order differential equation into *n* easy-to-solve, first-order differential equations. Additionally, you saw in the last chapter that a third-order differential equation simplified to a second-order. On closer examination, it was found that the model could be further simplified to a first-order differential equation.

In this chapter, we are going to further investigate the concept of dominant roots. You are going to discover that the solutions to higher-order linear ordinary differential equations with constant coefficients are comprised of multiple combinations of solutions that look like the first- and second-order differential equation solutions you've already studied. Roots of the characteristic polynomial of an *nth*-order differential equation occur either as single real exponentials or in pairs of complex exponentials. So what you have already learned with regard to the solution of first- and second-order differential equations applies to the solution of higher-order differential equations. You will also learn that any higher-order differential equation can, in many respects, be approximated by a second-order differential equation, or at most a combination of a first-order and second-order differential equation. You will learn how to spot those roots of the characteristic polynomial that govern the response of the system.

Overall, what you will learn in this chapter are ways to boil down more complex systems so you can use your intuition and knowledge of the behavior of first- and second-order systems to get engineering solutions to your design problems.

8.2 Constructing and Analyzing Higher-Order Math Models

The techniques introduced in Chapter 3 for developing math models apply regardless of the complexity of the system you are attempting to model. However, for more complex systems, there is much more algebra involved in deriving the final math model and solving the differential equations. The likelihood that you will make a mistake will be very high, so you must carefully check your work as you proceed. You are also going to find that the path-vertex-elemental equation approach becomes *very* tedious. There will be numerous equations, with numerous variables to eliminate. Nevertheless, this is still a fundamental, grass-roots approach to developing math models and it should be used at least to draw the block diagram of the system.

Here are some general rules to follow when preparing a higher-order, or for that matter any order, math model:

(1) Clearly define which variable is going to be the output and which is going to be the input.

(2) Each time you eliminate an unwanted variable, by substitution of one equation into other equations, check units of the resulting equations. Remember that each additive or subtractive term in an equation *must* have the same units.

(3) Perform limit case checks. As you saw in Chapter 6, you can often check the reasonableness of your answers by letting parameters go to zero. Often this causes the model to reduce to a simpler model for which you already have a solution. Similarly, letting parameters go to infinity can lead to a simplified model that is easy to check.

(4) When the system being studied has only passive elements (as is the case for all systems studied in this book), putting the differential equation in standard form (output variable on the left and input variable on the right) must result in all terms on both sides of the equation having positive signs. (This is a property of systems containing only passive elements.)

(5) Derive the math model using two or more techniques. (Quite often this will pinpoint problem areas.)

(6) Build symbolic diagrams (circuit diagrams and block diagrams) as you proceed with the development of the model.

(7) Check the order of the math model with the number of storage elements. The order of the math model cannot exceed the number of energy storage elements. However, the order of the math model *can* be less than the number of storage elements. This can occur when there are energy storage elements in parallel or in series. (This may be easier to visualize with mechanical

233

systems. Think of a spring with a small mass hanging from it. If you add another mass just under the first one, both masses act as one mass. But if you put the second mass on the other end of the spring, you have two separate energy sources.)

(8) Check the order of the left side (output variable side) of the differential equation. Its order cannot exceed the order of the right side (input variable side).

Third-Order Math Models
Model Construction (Case 1)

Take a look at the two circuits shown in Figure 8.1. The circuit on the left, consisting of R_1 and C_1, was studied in great detail in Chapters 3 and 4. The circuit on the right, consisting of R_2, C_2, and L_2, was studied in similar detail in Chapter 6. I've listed the math models of these two circuits showing V_a as the output of the circuit on the left and as the input to the circuit on the right. Let's combine the two circuits as shown in Figure 8.2 and prepare a math model with V_o as the output and V_i as the input.

You might be tempted to derive the math model by simply multiplying the two transfer functions for the circuits given in Figure 8.1. Unfortunately, this would not give you the right math model. That's because the math model for the R_1–C_1 circuit was developed using the assumption that *no* current flowed out of the circuit as indicated in Figure 8.1. If the R_2–C_2–L_2 circuit is connected to the R_1–C_1 circuit, it will draw current from the R_1–C_1 circuit and negate the no-current assumption. Drawing current from a circuit is often referred to as *loading the*

circuit. In this case, the R_2–C_2–L_2 circuit loads the R_1–C_1 circuit.

However, even though it won't lead to a correct model, let's start the investigation of this circuit by first multiplying the two transfer functions to create an approximate math model anyway. That is,

$$\frac{V_o}{V_i} = \frac{V_a}{V_i} \times \frac{V_o}{V_a}$$

$$= \left(\frac{1}{\tau_1 D + 1}\right) \times \left(\frac{1}{\frac{1}{\omega_n^2}D^2 + \frac{2\zeta}{\omega_n}D + 1}\right)$$

$$\frac{V_o}{V_i} = \frac{1}{\frac{\tau_1}{\omega_n^2}D^3 + \left(\tau_1\frac{2\zeta}{\omega_n} + \frac{1}{\omega_n^2}\right)D^2 + \left(\tau_1 + \frac{2\zeta}{\omega_n}\right)D + 1}$$

If we carry out the indicated operations, then the math model is also given by

$$\frac{\tau_1}{\omega_n^2}\frac{d^3V_o}{dt^3} + \left(\frac{2\zeta}{\omega_n}\tau_1 + \frac{1}{\omega_n^2}\right)\frac{d^2V_o}{dt^2}$$

$$+ \left(\tau_1 + \frac{2\zeta}{\omega_n}\right)\frac{dV_o}{dt} + V_o = V_i$$

or

$$R_1C_1L_2C_2\frac{d^3V_o}{dt^3} + \left(R_1C_1R_2C_2 + L_2C_2\right)\frac{d^2V_o}{dt^2}$$

$$+ \left(R_1C_1 + R_2C_2\right)\frac{dV_o}{dt} + V_o = V_i$$

$$(8.1)$$

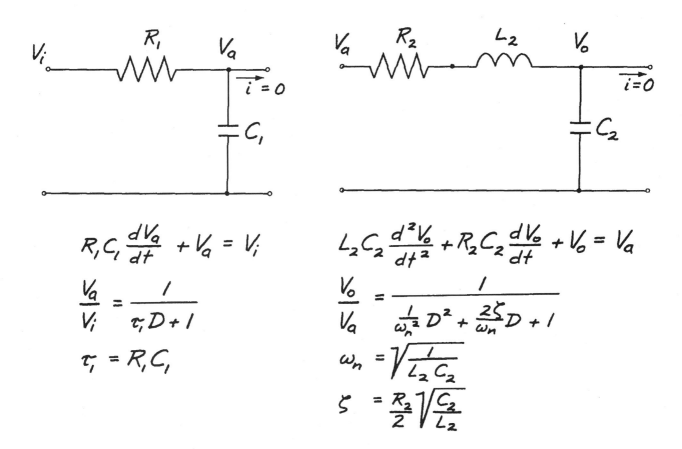

Figure 8.1. Two previously studied circuits.

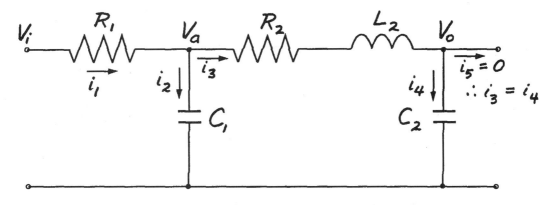

Figure 8.2. Circuits shown in Figure 8.1 connected together.

This is a third-order linear ordinary differential equation with constant coefficients.

Now let's develop the correct math model. Refer to Figure 8.2 and write the following equations:

$$V_a = V_i - R_1 i_1 \qquad (8.2)$$

$$V_o = V_a - R_2 i_3 - L_2 \frac{di_3}{dt} \qquad (8.3)$$

$$i_2 = C_1 \frac{dV_a}{dt} \qquad (8.4)$$

$$i_4 = C_2 \frac{dV_o}{dt} = i_3 \qquad (8.5)$$

$$i_1 = i_2 + i_3 \qquad (8.6)$$

Eliminate i_2 and i_3 by substituting (8.4) and (8.5) into (8.3) and (8.6). The circuit equations now consist of

$$V_a = V_i - R_1 i_1 \qquad \begin{matrix}(8.2)\\ \text{repeated}\end{matrix}$$

$$i_1 = C_1 \frac{dV_a}{dt} + C_2 \frac{dV_o}{dt} \qquad (8.7)$$

$$V_o = V_a - R_2 C_2 \frac{dV_o}{dt} - L_2 C_2 \frac{d^2 V_o}{dt^2} \qquad (8.8)$$

Eliminate i_1 by substituting (8.7) into (8.2). The circuit equations now consist of

$$V_a = V_i - R_1 C_1 \frac{dV_a}{dt} - R_1 C_2 \frac{dV_o}{dt} \qquad (8.9)$$

$$V_o = V_a - R_2 C_2 \frac{dV_o}{dt} - L_2 C_2 \frac{d^2 V_o}{dt^2} \qquad \begin{matrix}(8.8)\\ \text{repeated}\end{matrix}$$

Rearrange these two equations in the standard form:

$$R_1 C_1 \frac{dV_a}{dt} + V_a = V_i - R_1 C_2 \frac{dV_o}{dt} \qquad \begin{matrix}(8.9)\\ \text{rearranged}\end{matrix}$$

$$L_2 C_2 \frac{d^2 V_o}{dt^2} + R_2 C_2 \frac{dV_o}{dt} + V_o = V_a \qquad \begin{matrix}(8.8)\\ \text{rearranged}\end{matrix}$$

You'll recognize (8.8) as the math model we developed in Chapter 6 for the R_2–C_2–L_2 circuit. It has not changed because there is no current being drawn from the circuit at V_o (that is, $i_5 = 0$). You'll also recognize (8.9) as the math model we developed in Chapter 3 for the R_1–C_1 circuit, but with an added term $R_1 C_2 (dV_o/dt)$. If you look at (8.6) and (8.7) you'll see that this new term is due to the presence of current i_3. If i_3 were zero, this new term would disappear.

Let's eliminate V_a in (8.8) using (8.9). This can most easily be done by first placing (8.8) and (8.9) in operational notation form. That is,

$$V_a = \frac{V_i - R_1 C_2 D V_o}{(R_1 C_1 D + 1)} \qquad \begin{matrix}(8.9)\\ \text{operational}\\ \text{form}\end{matrix}$$

and

$$V_o = \frac{1}{L_2 C_2 D^2 + R_2 C_2 D + 1} V_a \qquad \begin{matrix}(8.8)\\ \text{operational}\\ \text{form}\end{matrix}$$

Now eliminate V_a and get

$$V_o = \frac{1}{L_2 C_2 D^2 + R_2 C_2 D + 1} \times \frac{V_i - R_1 C_2 D V_o}{(R_1 C_1 D + 1)}$$

Carry out the multiplication of the two polynomials and rearrange terms into the standard form:

$$R_1 C_1 L_2 C_2 \frac{d^3 V_o}{dt^3} + \left(R_1 C_1 R_2 C_2 + L_2 C_2\right)\frac{d^2 V_o}{dt^2}$$

$$+ \left(R_1 C_1 + R_2 C_2 + R_1 C_2\right)\frac{dV_o}{dt}$$

$$+ V_o = V_i \qquad (8.10)$$

Now let's compare the exact math model given by equation (8.10) with the approximation math model given by equation (8.1). Interestingly, the two equations are almost identical except for the addition of the term $R_1 C_2$ to the coefficient of the dV_o/dt term. You'll find that this is usually the case when developing math models by piecing together circuit components. A rough approximation of the combined circuit can be obtained by multiplying together transfer functions of the circuit components that were developed, assuming no loading of the output of these components. If one component circuit does not load the other too much, then multiplying the two transfer functions together gives a very good approximation math model of the combined circuit.

I went through the transfer function multiplication approximation process for an important reason. I wanted you to see that a third-order math model resulted from a combination of a first- and a second-order model. The third-order operator polynomial is simply the product of a first-order operator poly-

nomial and a second-order operator polynomial. Consequently, you can expect the solution to the third-order model to look like the combination of the solutions to the constituent first- and second-order models. We'll delve deeper into this in a moment.

Model Construction (Case 2)

A third-order model can also result from the combination of three first-order models. Figure 8.3 shows three R–C circuits ganged together. The rough approximation math model between V_o and V_i, obtained by multiplying the three transfer functions together, reveals that the true math model will be of third-order.

That is,

$$\frac{V_a}{V_i} = \frac{1}{R_1 C_1 D + 1} = \frac{1}{\tau_1 D + 1} \qquad (8.11a)$$

$$\frac{V_b}{V_a} = \frac{1}{R_2 C_2 D + 1} = \frac{1}{\tau_2 D + 1} \qquad (8.11b)$$

$$\frac{V_o}{V_b} = \frac{1}{R_3 C_3 D + 1} = \frac{1}{\tau_3 D + 1} \qquad (8.11c)$$

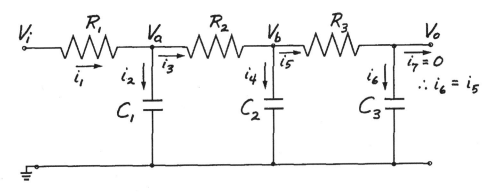

Figure 8.3. Three identical RC circuits ganged together.

Then as a first approximation

$$\frac{V_o}{V_i} = \frac{V_a}{V_i} \times \frac{V_b}{V_a} \times \frac{V_o}{V_b}$$

Carrying out the multiplication gives the approximate transfer function

$$\frac{V_o}{V_i} = 1 \ / \ [\tau_1\tau_2\tau_3 D^3 + (\tau_1\tau_2 + \tau_1\tau_3 + \tau_2\tau_3)D^2$$

$$+ (\tau_1 + \tau_2 + \tau_3)D + 1] \quad (8.12)$$

The correct transfer function (derived as shown in Figure 8.4) is

$$\frac{V_o}{V_i} = 1 \ / \ [\tau_1\tau_2\tau_3 D^3$$

$$+ (\tau_1\tau_3 + \tau_2\tau_3 + \tau_{12}\tau_3 + \tau_1\tau_2 + \tau_1\tau_{23})D^2$$

$$+ (\tau_1 + \tau_2 + \tau_3 + \tau_{12} + \tau_{13} + \tau_{23})D + 1]$$

$$(8.13)$$

where τ_{12}, τ_{13}, and τ_{23}, equals R_1C_2, R_1C_3, and R_2C_3, respectively. Note that (8.13) is of similar form to the approximation model, but with additional terms added to the coefficients due to one circuit loading the next. Here again, you can expect the solution of this third-order math model to look like the combination of solutions to three first-order models.

Model Analysis (Case 2)

We'll begin the analysis of a third-order math model starting with the circuit shown in Figure 8.3. First we'll obtain a numerical solution to the step response and then we'll obtain the exact solution. By now you should be able to numerically solve the three first-order differen-

tial equations given by equations (8.11a), (8.11b), and (8.11c) in Figure 8.3 without any coaching. Try to do this yourself and solve the equations for a step response of V_i equal to 10 volts with $R_1 = R_2 = R_3 = 1$ and $C_1 = C_2 = C_3 = 1$ and $V_a = V_b = V_o = 0$.

Figure 8.5 shows the step response obtained using a time step of 0.05 seconds. I've labeled the voltage at each of the junctions so you can compare these to the step response we obtained in Chapter 4. The algorithm I used in my spreadsheet is given in Table 8.1. Also, to help you understand the circuit and the step response better, I prepared a block diagram of the circuit in Figure 8.6.

The V_o response in Figure 8.5 looks a little like the step response of a first-order differential equation. The only difference is the initial slope of the response is zero. It's easy to see why this is, using the block diagram. There are three integrators between the input V_i and the output V_o. It takes time for the step change in the input voltage to have an effect on the output. However, there is only one integrator between the input voltage and the voltage V_a. That is why V_a has an initial slope (equal to $(V_i - V_a)/R_1$) and looks almost exactly like the step response of a first-order math model. You can also see this clearly in the spreadsheet simulation.

Compare the V_a response to the step response shown in Figure 4.2. Notice the step response in Figure 4.2 is much faster. That's because the loading caused by the second first-order circuit slows things down. You can see the effect more clearly in the block diagram. Current i_3, the load current being drawn from the R_1–C_1 circuit, is fed back and subtracted from the current i_1. This reduces

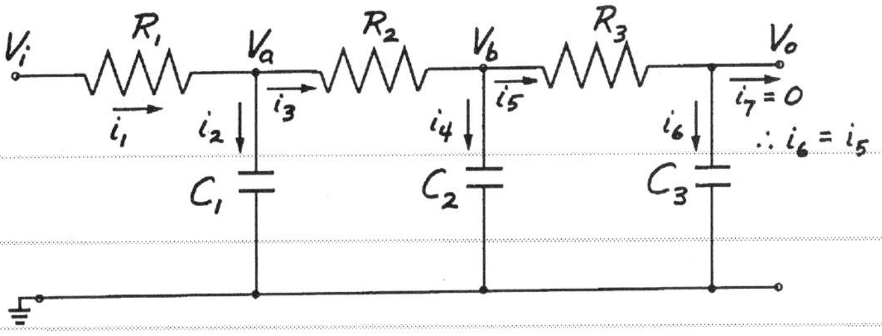

$$V_a = V_i - R_1 i_1 = V_i - R_1 C_1 \dot{V_a} - R_1 C_2 \dot{V_b} - R_1 C_3 \dot{V_o}$$

$$V_b = V_a - R_2 i_3 = V_a - R_2 C_2 \dot{V_b} - R_2 C_3 \dot{V_o}$$

$$V_o = V_b - R_3 i_5 = V_b - R_3 C_3 \dot{V_o}$$

$$i_2 = C_1 \dot{V_a} \qquad i_4 = C_2 \dot{V_b} \qquad i_5 = C_3 \dot{V_o}$$

$$i_1 = i_2 + i_3 = C_1 \dot{V_a} + C_2 \dot{V_b} + C_3 \dot{V_o}$$

$$i_3 = i_4 + i_5 = C_2 \dot{V_b} + C_3 \dot{V_o}$$

$$R_1 C_1 \dot{V_a} + V_a = V_i - R_1 C_2 \dot{V_b} - R_1 C_3 \dot{V_o} \;\rightarrow\; V_a = \frac{V_i - R_1 C_2 D V_b - R_1 C_3 D V_o}{R_1 C_1 D + 1}$$

$$R_2 C_2 \dot{V_b} + V_b = V_a - R_2 C_3 \dot{V_o} \;\rightarrow\; V_b = \frac{V_a - R_2 C_3 D V_o}{R_2 C_2 D + 1}$$

$$R_3 C_3 \dot{V_o} + V_o = V_b \;\rightarrow\; (R_3 C_3 D + 1) V_o = V_b$$

$$(R_2 C_2 D + 1) V_b = \frac{V_i - R_1 C_2 D V_b - R_1 C_3 D V_o}{R_1 C_1 D + 1} - R_2 C_3 D V_o$$

$$(R_1 C_1 D + 1)(R_2 C_2 D + 1) V_b = V_i - R_1 C_2 D V_b - R_1 C_3 D V_o - R_1 C_1 R_2 C_3 D^2 V_o - R_2 C_3 D V_o$$

$$[R_1 C_1 R_2 C_2 D^2 + (R_1 C_1 + R_2 C_2 + R_1 C_2)D + 1] V_b = V_i - R_1 C_1 R_2 C_3 D^2 V_o - (R_1 C_3 + R_2 C_3) D V_o$$

$$V_o (R_3 C_3 D + 1)[R_1 C_1 R_2 C_2 D^2 + (R_1 C_1 + R_2 C_2 + R_1 C_2)D + 1] = V_i - R_1 C_1 R_2 C_3 D^2 V_o - (R_1 C_3 + R_2 C_3) D V_o$$

$$V_o [R_1 C_1 R_2 C_2 R_3 C_3 D^3 + (R_1 C_1 R_3 C_3 + R_2 C_2 R_3 C_3 + R_1 C_2 R_3 C_3 + R_1 C_1 R_2 C_2 + R_1 C_1 R_2 C_3) D^2 +$$
$$+ (R_3 C_3 + R_1 C_1 + R_2 C_2 + R_1 C_2 + R_1 C_3 + R_2 C_3) D + 1] = V_i$$

$$V_o [\tau_1 \tau_2 \tau_3 D^3 + (\tau_1 \tau_3 + \tau_2 \tau_3 + \tau_{12} \tau_3 + \tau_1 \tau_2 + \tau_1 \tau_{23}) D^2 + (\tau_1 + \tau_2 + \tau_3 + \tau_{12} + \tau_{13} + \tau_{23}) D + 1] = V_i$$

Figure 8.4. Derivation of exact math model for circuit shown in Figure 8.3.

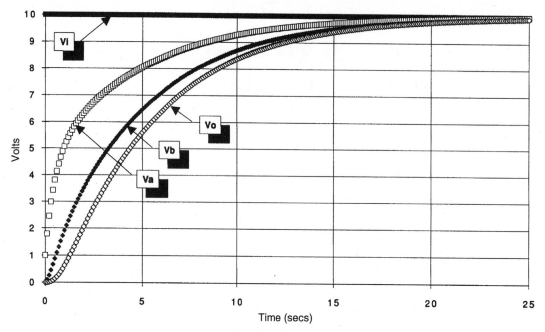

Figure 8.5. Step response of circuit in Figure 8.3.

Table 8.1.
Algorithm used in spreadsheet.

$$\dot{V}_a = \frac{1}{C_1}i_2 \qquad \Delta V_a = \dot{V}_a\,\Delta t \qquad V_a = V_a + \Delta V_a \qquad i_1 = \frac{1}{R_1}(V_i - V_a) \qquad i_2 = i_1 - i_3$$

$$\dot{V}_b = \frac{1}{C_2}i_4 \qquad \Delta V_b = \dot{V}_b\,\Delta t \qquad V_b = V_b + \Delta V_b \qquad i_3 = \frac{1}{R_3}(V_a - V_b) \qquad i_4 = i_3 - i_5$$

$$\dot{V}_c = \frac{1}{C_3}i_5 \qquad \Delta V_o = \dot{V}_o\,\Delta t \qquad V_o = V_o + \Delta V_o \qquad i_5 = \frac{1}{R_5}(V_b - V_o)$$

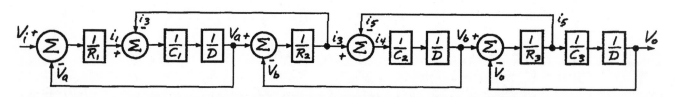

Figure 8.6. Block diagram of the circuit shown in Figure 8.3.

the current going into the first integrator. Compare this block diagram to the one given in Figure 3.10 for the same circuit without a load—there is no i_3 feedback in Figure 3.10 to slow the current into the capacitor. Study this circuit until you have a good feel for its response.

Now let's find the exact solution. Proceed as described in Chapters 4 and 6. That is, first obtain the homogeneous solution and then the particular solution. Add the two together to arrive at the complete solution.

The exact math model is given above by (8.13). This can be placed in our standard form by carrying out the indicated operations. That is,

$$\tau_1 \tau_2 \tau_3 \frac{d^3 V_o}{dt^3}$$

$$+ \left(\tau_1 \tau_3 + \tau_2 \tau_3 + \tau_{12} \tau_3 + \tau_1 \tau_2 + \tau_1 \tau_{23} \right) \frac{d^2 V_o}{dt^2}$$

$$+ \left(\tau_1 + \tau_2 + \tau_3 + \tau_{12} + \tau_{13} + \tau_{23} \right) \frac{dV_o}{dt} + V_o = V_i$$

Substituting values for the circuit components used in the numerical solution gives

$$\frac{d^3 V_o}{dt^3} + 5 \frac{d^2 V_o}{dt^2} + 6 \frac{dV_o}{dt} + V_o = V_i \tag{8.14}$$

The homogeneous solution is obtained by substituting the solution

$$\left(V_o \right)_H = A e^{-rt}$$

and its derivatives into (8.14). Carrying out the differentiation and making substitutions leads to the following third-order polynomial in r:

$$r^3 + 5r^2 + 6r + 1 = 0 \tag{8.15}$$

You may recall from algebra that a third-order polynomial must have three roots r_1, r_2, and r_3 such that

$$(r - r_1)(r - r_2)(r - r_3) = r^3 + 5r^2 + 6r + 1$$

As a consequence, the homogeneous solution has the form:

$$\left(V_o \right)_H = A_1 e^{-r_1 t} + A_2 e^{-r_2 t} + A_3 e^{-r_3 t} \tag{8.16}$$

We now need to find the roots of the characteristic polynomial given by (8.15). There are numerous ways to do this and I'll discuss some in this chapter. However, I want to warn you that finding the roots to a polynomial can be very tedious. I highly recommend that you get a computer program or a calculator that can find roots of the polynomial given its degree and coefficients. This will make your math-model–building life much more pleasant.

One of the easiest methods of finding roots of a polynomial is to plot the *value* of the polynomial as a function of r. The value of the polynomial must be zero at each root in order to satisfy (8.15). This will occur where the value of the polynomial crosses the r axis. Figure 8.7 shows the plot.

Ordinarily you would plot the value of the polynomial for positive and negative r. However, there are no positive values of r that cause the polynomial to evaluate to zero. The value of the polynomial is zero only at the following values of r:

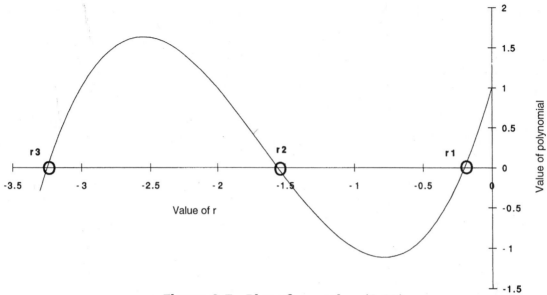

Figure 8.7. Plot of equation (8.15).

$$r_1 = -0.20$$

$$r_2 = -1.55$$

$$r_3 = -3.25$$

Using these values for the roots, the homogeneous solution becomes

$$(V_o)_H = A_1e^{-0.20t} + A_2e^{-1.55t} + A_3e^{-3.25t}$$
$$(8.17)$$

where the coefficients A_1, A_2, and A_3 are to be determined using the initial conditions.

Note that all of the roots in this solution are *real* and *negative*. This indicates that the solution is the sum of decaying exponentials. Had any of the roots been positive, the solution would grow without bound as time increased and the system would be unstable. A plot of the roots on the complex plane is shown in Figure 8.8. The root

at -0.20 is clearly nearest the origin and will dominate the solution because it has the longest time constant ($\tau = 1 / 0.20 = 5$ seconds).

The particular solution is found by substituting

$$(V_o)_P = k_0$$

into (8.14). All of the derivatives of this solution are zero, so after substitution we're left with

$$k_0 = V_{is}$$

which satisfies (8.14) for all time.

The total solution is obtained by adding together the homogeneous and particular solutions. That is,

$$V_o = V_{is} + A_1e^{-0.20t} + A_2e^{-1.55t} + A_3e^{-3.25t}$$
$$(8.18)$$

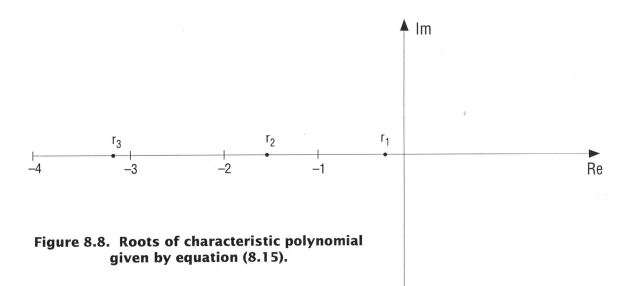

Figure 8.8. Roots of characteristic polynomial given by equation (8.15).

The coefficients A_1, A_2, and A_3 are obtained from the initial conditions at time $t = 0$. With the circuit initially at rest, these initial conditions are:

$$V_o = 0$$

$$\frac{dV_o}{dt} = 0$$

$$\frac{d^2V_o}{dt^2} = 0$$

Taking the indicated derivatives of (8.18) and setting the results to zero gives the following three equations for determining the coefficients:

$$+1.00A_1 + 1.00A_2 + 1.00A_3 = -V_{is}$$

$$-0.20A_1 - 1.55A_2 - 3.25A_3 = 0$$

$$-0.04A_1 - 2.40A_2 - 10.56A_3 = 0$$

These three equations have three unknowns, which means they can be solved by the process of variable elimination. Carrying out the algebra results in:

$$A_1 = -1.1766V_{is}$$

$$A_2 = +0.2228V_{is}$$

$$A_3 = -0.0462V_{is}$$

These coefficients can now be substituted back into (8.18), giving the complete solution as:

$$V_o = V_{is}\left(1 - 1.11766e^{-0.20t} + 0.2228e^{-1.55t} - 0.0462e^{-3.25t}\right)$$

$$(8.19)$$

Figure 8.9 shows the relative contribution of each term in (8.18) to the total solution. You can see that the term with the root equal to −0.20 dominates the solution. The other two terms die out quickly because their time constants are so small. Always keep this in mind: ROOTS IN THE COMPLEX PLANE NEAREST THE ORIGIN DOMINATE THE RESPONSE.

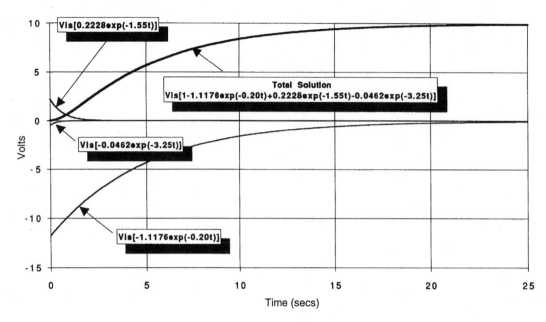

Figure 8.9. Step response of Figure 8.3 showing total solution and components (R1 = C1 = R2 = C2 = R3 = C3 =1).

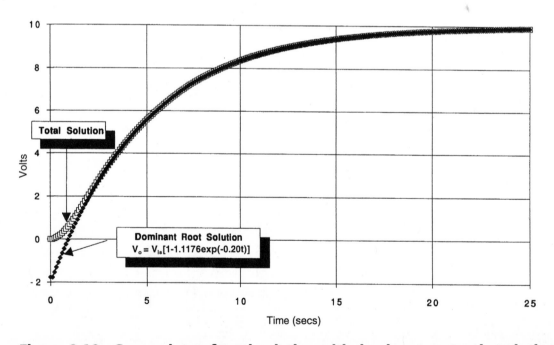

Figure 8.10. Comparison of total solution with dominant root only solution.

Figure 8.10 compares the exact total solution with an approximation using only the first term in (8.19). It's clear that, except for the initial part of the response, the approximate response using only the first term of the solution is very close to the exact solution. In essence, the third-order math model described by (8.14) is behaving approximately as if it were a first-order math model.

Let's now take a look at the frequency response of this circuit. You can proceed as explained in Chapter 6. That is, substitute $j\omega$ into the transfer function given by (8.13) and get

$$\frac{V_o}{V_i} = \frac{1}{(j\omega)^3 + 5(j\omega)^2 + 6(j\omega) + 1}$$

$$= \frac{1}{(1 - 5\omega^2) + j(6\omega - \omega^3)}$$

$$= \frac{1}{\sqrt{(1 - 5\omega^2)^2 + (6\omega - \omega^3)^2}} e^{-j\phi}$$

$$\tag{8.20}$$

where

$$\phi = \tan^{-1}\left(-\frac{6\omega - \omega^3}{1 - 5\omega^2}\right)$$

$$\tag{8.21}$$

The magnitude of the transfer function is plotted in Figure 8.11 and the phase angle is plotted in Figure 8.12. Notice that as the frequency increases, the amplitude ratio decreases at an increasing rate. This is due to the fact that there are three first-order frequency responses present in this system.

It's easy to construct an approximate frequency response of a transfer function when you have the roots of the characteristic polynomial.

I have indicated how this is done on Figure 8.11. For each root you draw the low-frequency approximation (which is one for each root in this example) up to the break frequency. Then you draw in the high-frequency approximation starting at the break frequency with a slope of –20 db per decade. For the first root the high-frequency approximation is $0.20/\omega$, for the second it's $1.55/\omega$, and for the third it's $3.25/\omega$. As the frequency increases, you keep multiplying the approximations together to get the approximate response. This is a very handy way of estimating the frequency response of complex systems. It also comes in handy estimating the make-up of an engineering system given only its frequency response.

Model Analysis (Case 1)

Now we'll analyze the third-order circuit whose schematic is given in Figure 8.1 and whose math model is given by (8.10). Take the same approach as before and obtain a numerical solution for V_o using a step input of V_i = 10 volts. Use $R_1 = C_1 = 1$, $C_2 = L_2 = 1$, $R_2 = 0.6$, and let the initial conditions be zero. These values are identical to those used in Chapter 6, Figure 6.10 for the R_2–L_2–C_2 circuit. After investigating the circuit using numerical solutions, we'll obtain the exact solution and investigate the circuit further.

Figure 8.13 shows the step response I obtained with a spreadsheet simulation using a time step of 0.05 seconds. I've labeled the voltage at the junctions so you can again compare the step response of the R_1–C_1 circuit to that we obtained in Chapter 4 and the step response of

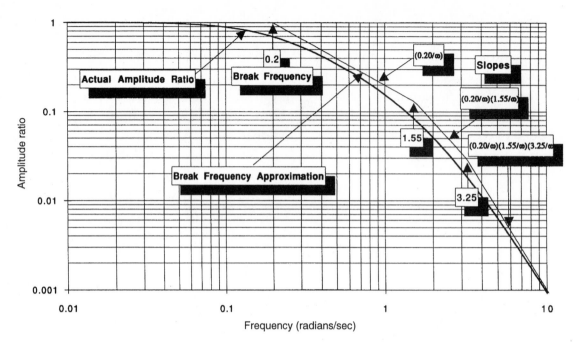

Figure 8.11. Frequency response of circuit shown in Figure 8.3.

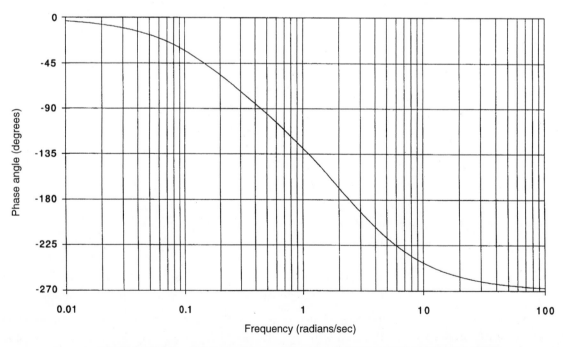

Figure 8.12. Frequency response of circuit in Figure 8.3.

the R_2–L_2–C_2 circuit in Chapter 6. I've also prepared a block diagram of the circuit in Figure 8.14.

Compare the V_o response of Figure 8.13 with the step response given in Figure 6.10. The R_1–C_1 circuit is obviously having a dampening effect on the output of the R_2–L_2–C_2 circuit. You can see that it does by looking at the V_a response. It is this voltage that the R_2–L_2–C_2 circuit is responding to.

Note the shape of the V_a response. Clearly the oscillatory behavior of the R_2–L_2–C_2 circuit is affecting the R_1–C_1 circuit. Carefully review the block diagram in Figure 8.14 and then compare it with block diagrams given in Figure 3.9 and Figure 6.3c. Notice that it is the feedback of i_3 into the R_1–C_1 circuit that is causing the oscillatory shape of the V_a response.

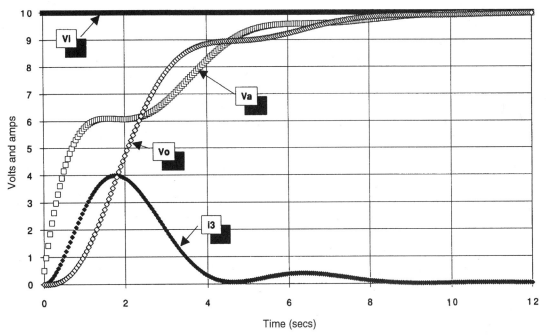

Figure 8.13. Step response of circuit in Figure 8.1 (R1 = C1 = L2 = C2 = 1, R2 = 0.6).

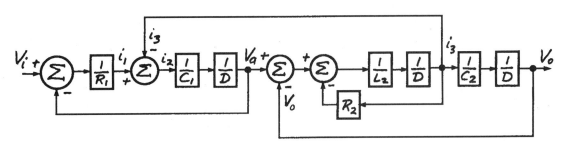

Figure 8.14. Block diagram of circuit shown in Figure 8.1.

If it is the V_a response that the R_2–L_2–C_2 circuit is responding to, then making the time constant of the R_1–C_1 circuit small should make the V_a response faster and the V_o response approach that shown in Figure 6.10. To check this, I changed R_1 to 0.1 ohms in my spreadsheet simulation. That gives a time constant for the R_1–C_1 circuit of 0.1 seconds—that is, 1/10th of what it was in Figure 8.13. My numerical solution results are shown in Figure 8.15. If you compare the V_o response in this figure with that in Figure 6.10, you can see that they are almost identical.

Try to visualize what is going on here. Look at the sketch I made in Figure 8.16 of the approximate location of the three roots in the complex plane. Two of the roots are complex conjugates and the other is real. You can figure this out because: (a) there must be three roots to the characteristic polynomial; and (b) the response is oscillatory. An oscillatory response means there is a pair of complex roots and complex roots always come in conjugate pairs. The real root in Figure 8.13 is probably quite close to the origin and is having a dominant effect on the V_o response. I decreased the time constant of the R_1–C_1 circuit in Figure 8.15, moved this root away from the origin, and made the complex roots dominant.

We can test this again by increasing the value of the R_1–C_1 circuit time constant. This should move the real root closer to the origin and make it even more dominant. Figure 8.17 shows the response I obtained with R_1 equal to 4. Now the response looks more like the response of a first-order system to a step change in input.

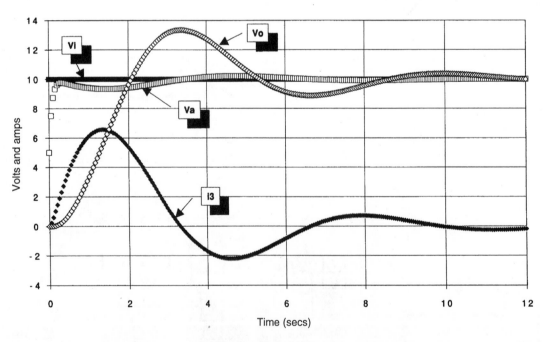

Figure 8.15. Step response of circuit in Figure 8.1
(R1 = 0.1, C1 = L2 = C2 = 1, R2 = 0.6).

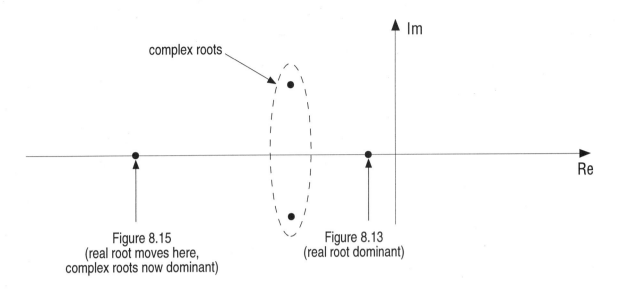

Figure 8.16. Approximate location of roots for circuit shown in Figure 8.1.

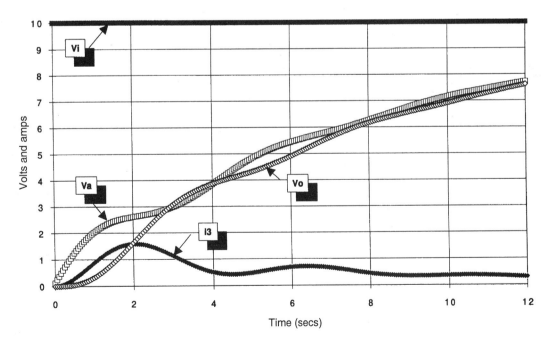

Figure 8.17. Step response of circuit in Figure 8.1
(R1 = 4, C1 = L2 = C2 = 1, R2 = 0.6).

Given that the R_1–C_1 circuit is having a damping effect on the V_o response to a step, we should be able to reduce the dampening in the R_2–L_2–C_2 circuit. In Figure 6.13, we studied the step response of the R_2–L_2–C_2 circuit with R_2 equal to zero. Figure 8.18 shows the step response with R_2 equal to zero and the R_1–C_1 circuit back to a 1-sec time constant. While we have made the V_o response more oscillatory, it is clear that the circuit has much more damping with the R_1–C_1 circuit than it did without it.

We could go on with this investigation using the numerical solution, and I strongly recommend that you set one up and experiment with it. For now, however, let's move on to finding the exact solution for the case $R_1 = C_1 = R_2 = C_2 = L_2 = 1$.

As before, the homogeneous solution is obtained by substituting the solution

$$(V_o)_H = Ae^{-rt}$$

into the differential equation (8.10). This leads to the requirement:

$$r^3 + 2r^2 + 3r + 1 = 0 \qquad (8.22)$$

We again seek the roots r_1, r_2, and r_3 such that

$$(r - r_1)(r - r_2)(r - r_3) = r^3 + 2r^2 + 3r + 1$$

The homogeneous solution will again have the form:

$$(V_o)_H = A_1 e^{-r_1 t} + A_2 e^{-r_2 t} + A_3 e^{-r_3 t} \quad (8.23)$$

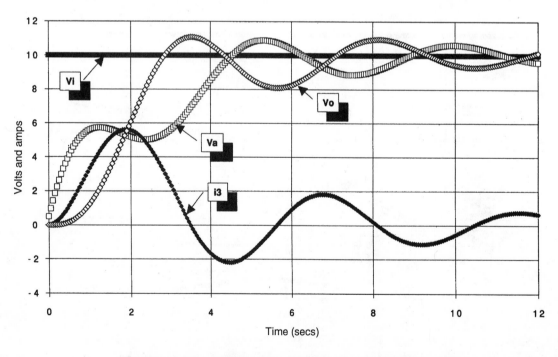

Figure 8.18. Step response of circuit in Figure 8.1 (R1 = C1 = L2 = C2 = 1, R2 = 0).

Let's try again to find the roots of the polynomial by plotting the value of (8.22) as a function of r. Figure 8.19 shows the plot, but there appears to be only one root at –0.43. Since we know there are two more roots, then the other two roots must be complex. How do we go about finding those roots?

You may recall from algebra that the order of a polynomial can be reduced by one when one of the roots are known. You do this by a process of polynomials division called *synthetic division*. Synthetic division is really simple. It is just the reverse of multiplying one polynomial by another. Since we already have one root, we know that $(r + 0.43)$ is a root of (8.22). The other two roots must be contained in a quadratic equation such that

$$(r + 0.43)(ar^2 + br + 1) = r^3 + 2r^2 + 3r + 1$$

We therefore want to find $ar^2 + br + 1$ by dividing $(r + 0.43)$ into (8.22). Proceed as follows:

First set up for synthetic division by placing the divisor polynomial on the left side and the dividend polynomial on the inside, as shown below

$$r + 0.43 \overline{\smash{)}\, r^3 + 2r^2 + 3r + 1}$$

Now divide r into r^3 and put the result on top of the division symbol. Multiply this partial quotient and the divisor and write the results under the dividend. Subtract the result from the dividend as shown below to obtain a new dividend:

$$
\begin{array}{r}
r^2 \\
r + 0.43 \overline{\smash{)}\, r^3 + 2r^2 + 3r + 1} \\
\underline{r^3 + .43r^2} \\
1.57r^2 + 3r + 1
\end{array}
$$

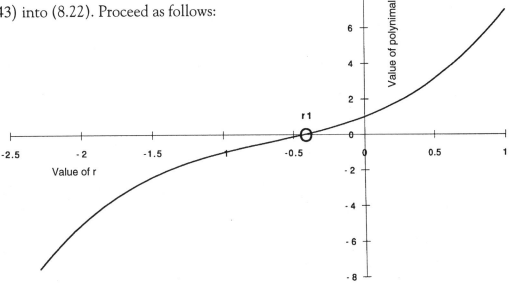

Figure 8.19. Plot of equation (8.16).

Now divide r into r^2 and put the result on top of the division symbol. Multiply and carry the results down another line as before:

$$
\begin{array}{r}
r^2 + 1.57r \\
r + 0.43 \overline{\smash{\big)}\, r^3 + 2r^2 + 3r + 1} \\
\underline{r^3 + .43r^2} \\
1.57r^2 + 3r + 1 \\
\underline{1.57r^2 + .675r} \\
2.325r + 1
\end{array}
$$

Now divide r into r, put the result on top of the division symbol, multiply and carry the results down another line:

$$
\begin{array}{r}
r^2 + 1.57r + 2.325 \\
r + 0.43 \overline{\smash{\big)}\, r^3 + 2r^2 + 3r + 1} \\
\underline{r^3 + .43r^2} \\
1.57r^2 + 3r + 1 \\
\underline{1.57r^2 + .675r} \\
2.325r + 1 \\
\underline{2.325r + 1} \\
0
\end{array}
$$

The remainder is now zero. We have found the quadratic polynomial

$$
r^2 + 1.57r + 2.325
$$

The above quadratic can easily be solved using the quadratic equation as we did in Chapter 6. The results are:

$$
r_2 = -0.785 + j1.309
$$
$$
r_3 = -0.785 - j1.309
$$

We now complete the homogeneous solution as

$$
\left(V_o \right)_H = A_1 e^{-0.43t}
$$
$$
+ A_2 e^{(-0.785 + j1.309)t}
$$
$$
+ A_2^* e^{(-0.785 - j1.309)t} \qquad (8.24)
$$

Note that I have taken advantage of the knowledge that when a pair of roots are complex, the coefficients are also complex conjugates. Equivalently, we can write (8.24) in the real form

$$
\left(V_o \right)_H = A_1 e^{-0.43t} + e^{-0.785t}
$$
$$
\times \left[B_1 \sin(1.309t) + B_2 \cos(1.309t) \right]
$$
$$
(8.25)
$$

as explained in Chapter 6.

Take note that all of the real parts of the complex roots in this solution are *negative*. This indicates that the solution is the sum of *decaying* exponentials with an oscillatory part. Once again the importance of the negative real parts should be noted. Had they been even slightly positive, the solution would increase without bound as time increased. A plot of the roots on the complex plane is shown in Figure 8.20. The root at −0.43 is clearly closer to the origin than are the complex roots. This is what we expected from our numerical solution.

The particular solution is found by substituting

$$
\left(V_o \right)_P = k_0
$$

When this solution is substituted into (8.11), all of the derivatives are zero and we're left with

$$
k_0 = V_{is}
$$

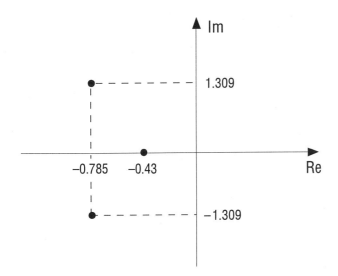

Figure 8.20. Roots of characteristic polynomial given by equation (8.22).

That is, the circuit is entirely discharged at $t = 0$. Taking the indicated derivatives of (8.26) and evaluating the result at $t = 0$ gives the following three equations for determining the coefficients:

$$+1.000A_1 + 1.000B_2 = -Vis$$

$$-0.430A_1 + 1.309B_1 - 0.785B_2 = 0$$

$$-0.185A_1 - 2.055B_1 - 1.097B_2 = 0$$

Since there are three equations and three unknowns, we can solve for the coefficients by elimination with the results:

$$A_1 = -1.2426V_{is}$$

$$B_1 = -0.2633V_{is}$$

$$B_2 = +0.2426V_{is}$$

This solution satisfies (8.11) for all time.

The total solution is obtained by summing the homogeneous and particular solutions, giving:

$$V_o = V_{is} + A_1 e^{-0.43t} + e^{-0.785t}$$
$$\times \left[B_1 \sin(1.309t) + B_2 \cos(1.309t) \right] \quad (8.26)$$

We obtain the coefficients A_1, B_1, and B_2 from the initial conditions at time $t = 0$. We'll use for the initial conditions:

$$V_o = 0$$

$$\frac{dV_o}{dt} = 0$$

$$\frac{d^2V_o}{dt^2} = 0$$

Substituting these coefficients back into (8.26) gives the complete solution as:

$$V_o = V_{is}\{1 - 1.2426e^{-0.43t} + e^{-0.785t}$$
$$\times[-0.2633\sin(1.309t)$$
$$+ 0.2426\cos(1.309t)]\} \quad (8.27)$$

Figure 8.21 shows the relative contribution of each term in (8.27) to the total solution. You can see that the term with the root equal to −0.43 dominates the solution. The other two terms die out quickly because the exponential term associated with the sine and cosine terms dies out quickly. So, once again, you see that ROOTS IN THE COMPLEX PLANE NEAREST THE ORIGIN DOMINATE THE RESPONSE EVEN WHEN THERE ARE COMPLEX ROOTS.

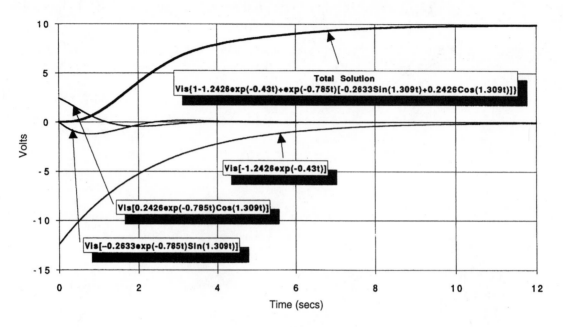

Figure 8.21. Step response of circuit in Figure 8.1 showing total solution and components (R1 = C1 = R2 = L2 = C2 = 1).

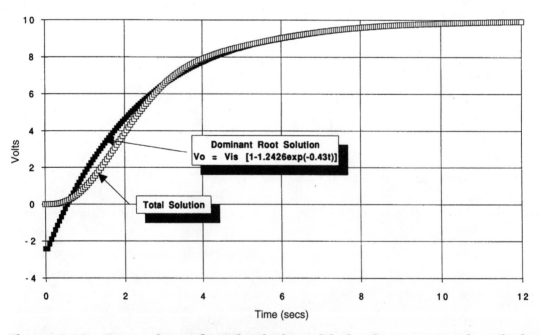

Figure 8.22. Comparison of total solution with dominant root only solution.

Figure 8.22 compares the exact total solution with an approximation using only the first term in (8.27). It's pretty clear that except for the initial response, the approximate response using only the first term of the solution is very close to the exact solution. IN ESSENCE, THE THIRD-ORDER MATH MODEL DESCRIBED BY (8.11) IS BEHAVING APPROXIMATELY AS IF IT WERE A FIRST-ORDER MODEL.

Let's now take a look at the frequency response of this circuit. We'll proceed as we did above for the other circuit. That is, substitute $j\omega$ into the transfer function derived from (8.10). That gives,

$$\frac{V_o}{V_i} = \frac{1}{(j\omega)^3 + 2(j\omega)^2 + 3(j\omega) + 1}$$

$$= \frac{1}{(1 - 2\omega^2) + j(3\omega - \omega^3)}$$

$$= \frac{1}{(1 - 2\omega^2) + j(3\omega - \omega^3)} \qquad (8.28)$$

where

$$\phi = \tan^{-1}\left(-\frac{3\omega - \omega^3}{1 - 2\omega^2}\right) \qquad (8.29)$$

The magnitude of the transfer function is plotted in Figure 8.23 and the phase angle is plotted in Figure 8.24. Notice that, as the frequency increases, the amplitude ratio decreases at an increasing rate. This is due to the fact that there are three first-order frequency responses present in this system.

We can construct an approximate frequency response of the transfer function just as we did

before. I have indicated how this is done on Figure 8.23. For the single real root, draw the low frequency approximation (which is one) up to the break frequency. Then draw in the high-frequency approximation starting at the break frequency with a slope of –20 db per decade, as we discussed earlier in the chapter. For the real root, the high frequency approximation is $0.43/\omega$. For the complex roots, we follow a similar procedure except we treat the roots together rather than individually. Draw the low-frequency response (which is one) up to the natural frequency. Then draw in the high-frequency approximation with a slope of –40 db per decade. As before, as the frequency increases, you simply keep multiplying the approximations together to get the approximate response.

Fourth-Order Math Models

Fourth- and higher-order math models will result when you model engineering systems that have four or more independent energy storage elements. For example, multiplying two second-order transfer functions (each having two independent energy storage elements) will result in a fourth-order transfer function. Similarly, multiplying five first-order transfer functions (each with one energy storage element) will result in a fifth-order transfer function. You can construct these higher-order math models by proceeding exactly as we did above with the third-order model—there will just be a lot more algebra!

I'm not going to go any further with this development of higher-order math models because I don't want to lose you in the algebra, and

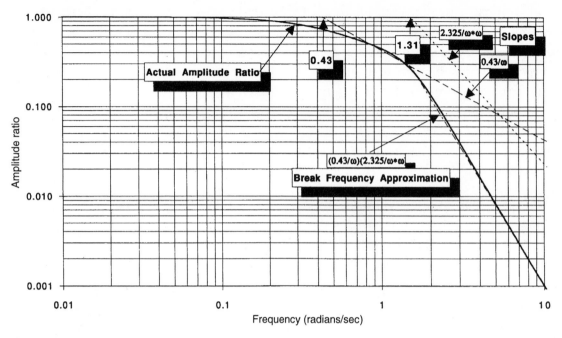

Figure 8.23. Amplitude ratio frequency response of circuit shown in Figure 8.1.

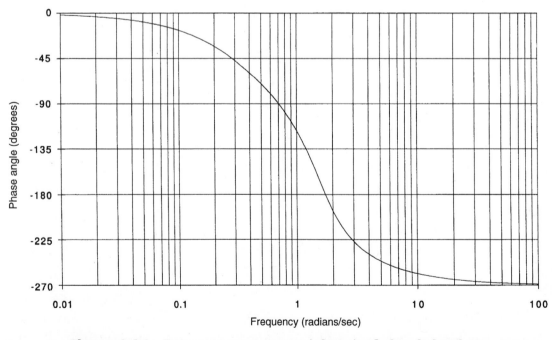

Figure 8.24. Frequency response (phase) of circuit in Figure 8.1.

because I'm nearing the end of this book. However, keep in mind that higher-order models may have real roots that behave exactly like the real root in a first-order math model, and/or complex roots that behave exactly like the complex roots in a lightly damped second-order math model. The root (or roots, if they are complex) nearest the origin in the complex plane will dominate the time domain solution. Because of this, most of the time you are going to find that, regardless of the order of the math model you come up with, there is a first- or second-order math model that can be obtained from the dominant roots that behaves almost exactly like your higher-order math model. Given the approximate nature of building math models of engineering systems to begin with, a simplified math model that you can understand and can easily relate to reality is far better than one that is so complex that you get lost in the world of mathematics. Knowing how to simplify math models to the bare essentials is really what makes great engineers. It's the stuff that engineering intuition is made of!

Appendix
A

Engineering Calculus
and Other Basic
Math Modeling Tools

A.1 Introduction

Even though calculus is taught at the high-school level today, many technically oriented people (engineers among them) still avoid it like the plague. For some reason, many of us—maybe because of past bad experiences with math—feel uncomfortable with calculus. The problem is made worse by the fact that it seems it is always taught by people who know little about engineering. However, whether you like hearing this or not, I have to say it: A good engineer *needs* to be comfortable with calculus. If you don't master this essential tool at some point, you will limit your career.

This appendix is an attempt to present a low-B.S, fog-free review of the essentials of calculus from an engineering systems point of view. I've tried to keep fancy words and theorems to a minimum. I use a familiar engineering system—your automobile—to introduce the concepts of differential and integral calculus.

A.2 Differential Calculus

One of the first helpful things to realize about calculus is that you deal with it every day. If you drive a car, every time you use the gas pedal you are directly applying calculus as you control the car's speed. Assume your car is at a standstill and you tramp down on the gas pedal. Refer to Figure A.1 and let's dig into calculus.

At some time t you will have traveled a distance x from the starting point. We say x is a function of time and write it

using symbols like $x(t)$, or $x = f(t)$, where $f(\ )$ means "a function of."

During a small time interval Δt, your car will travel a distance Δx. This incremental change in distance can be expressed in equation form as:

$$\Delta x = x(t + \Delta t) - x(t) \qquad (A.1)$$

That is, the change in distance is equal to the future position $x(t+\Delta t)$ minus the present position $x(t)$.

If both sides of (A.1) are divided by the time interval Δt then

$$\frac{\Delta x}{\Delta t} = \frac{x(t + \Delta t) - x(t)}{\Delta t} \qquad (A.2)$$

If Δx is measured in feet and Δt in seconds, the units of $\Delta x / \Delta t$ are feet per second—otherwise known as velocity. Equation (A.2) is an approximation of the velocity of your car. I'll use the symbol v to indicate velocity.

Let's work with an example. Say the distance your car is away from its starting point is given by the function

$$x(t) = 5t^2 \qquad (A.3)$$

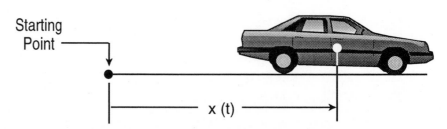

Starting Point

x (t)

Figure A.1. At time *t* the automobile will have travelled a distance *x(t)*.

and you want to determine its velocity two seconds after you tramped down on the accelerator (that is at $t = 2$ sec). Using equation (A.2) we write

$$\frac{\Delta x}{\Delta t} = \frac{x(t + \Delta t) - x(t)}{\Delta t} = \frac{5(t + \Delta t)^2 - 5t^2}{\Delta t} \quad (A.4)$$

If we now expand equation (A.4) we get

$$\frac{\Delta x}{\Delta t} = \frac{5\left(t^2 + 2 \cdot \Delta t \cdot t + \Delta t^2\right) - 5t^2}{\Delta t}$$
$$= \frac{10 \cdot \Delta t \cdot t + 5\Delta t^2}{\Delta t}$$

or

$$\frac{\Delta x}{\Delta t} = 10t + 5\Delta t \quad (A.5)$$

Now let's apply equation (A.5) to determine the velocity at $t = 2$ using various values for the time increment Δt as shown in Table A.1. You can see from this table that as Δt gets smaller, the velocity approximation $\Delta x / \Delta t$ is approaching 20 ft/sec.

Table A.1.

Δt (seconds)	v(ft/sec) at t = 2 seconds
1.0000	10 x 2 + 5 x (1.0000)^2 = 25.00000000
0.5000	10 x 2 + 5 x (0.5000)^2 = 21.25000000
0.2500	10 x 2 + 5 x (0.2500)^2 = 20.31250000
0.1000	10 x 2 + 5 x (0.1000)^2 = 20.05000000
0.0100	10 x 2 + 5 x (0.0100)^2 = 20.00050000
0.0010	10 x 2 + 5 x (0.0010)^2 = 20.00000500
0.0001	10 x 2 + 5 x (0.0001)^2 = 20.00000005

From equation (A.5) we can see that

$$\frac{\Delta x}{\Delta t} = 10t \qquad \text{as } \Delta t \to 0 \qquad (A.6)$$

So we go back to equation (A.2) and define the *derivative of a function x(t)* with respect to t as

$$\frac{dx}{dt} = \lim_{\Delta t \to 0} \frac{\Delta x}{\Delta t} = \lim_{\Delta t \to 0} \frac{x(t + \Delta t) - x(t)}{\Delta t}$$

$$(A.7)$$

As you just saw in the example, equation (A.7) can be used to derive the derivative of a function.

Figure A.2 shows a plot of equation A.3.

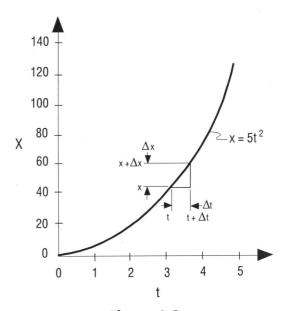

Figure A.2.

You can see that Δx divided by Δt is a slope. As Δt gets smaller and smaller, this slope approaches the derivative $\Delta x / \Delta t$. Therefore, the derivative of function $x(t)$ at the time t is equal to the slope of the function at time t.

Go back now to equation (A.6) and rewrite it as

$$v = \frac{dx}{dt} = 10t \qquad (A.8)$$

You can see that the velocity v is also a function of time. That is, $v = v(t)$. Since $v(t)$ is just another function of time, we can also take the derivative of this function. That is,

$$\frac{dv}{dt} = \lim_{\Delta t \to 0} \frac{\Delta v}{\Delta t} = \lim_{\Delta t \to 0} \frac{v(t + \Delta t) - v(t)}{\Delta t} \qquad (A.9)$$

or

$$\frac{dv}{dt} = \lim_{\Delta t \to 0} \frac{10(t + \Delta t) - 10t}{\Delta t} = \frac{10\Delta t}{\Delta t} = 10 \qquad (A.10)$$

In this case, the derivative of velocity dv/dt (which has the units of feet per second per second, or acceleration) is a constant. Now you know that your car is accelerating at 10 ft/sec², which of course you would feel on your back as you are pressed against the seat.

We twice differentiated the function describing the distance of your car from its starting point. This is called *double differentiation* and can be expressed as

$$\frac{d}{dt}\left(\frac{dx}{dt}\right) = \frac{d^2x}{dt^2} \qquad (A.11)$$

This is called taking the second derivative of a function.

We can also write derivatives in shorthand as

$$\frac{dx}{dt} = \dot{x} = v \qquad (A.12)$$

$$\frac{d^2x}{dt^2} = \ddot{x} = \dot{v} = a \qquad (A.13)$$

You can also use the *operator*

$$D = \frac{d}{dt} \qquad (A.14)$$

to express the derivative. This can be extremely handy. For example, equations (A.12) and (A.13) above can be expressed in operator notation form as follows

$$Dx = \frac{dx}{dt} = \dot{x} = v \qquad (A.15)$$

$$D^2x = \frac{d^2x}{dt^2} = \ddot{x} = \dot{v} = a \qquad (A.16)$$

Throughout this book block diagrams are used as an aid in building mathematical models. If $x(t)$ is an input or forcing function into the block below and the output is the derivative of the function, then the block must contain the differentiation operator. That is,

is the same as equation (A.15) and

is the same as equation (A.16). The operator D is often called a *differentiator* when used in block diagrams.

A differentiator can also be written for a digital computer. Listing A.1 is a "differentiator" BASIC computer program and Listing A.2 is a spreadsheet version. Carefully review these pro-

Listing A.1.
Digital computer differentiator.

```
5    REM BASIC DIFFERENTIATOR

10   DEF FN X(T) = 5 * T^2

20   INPUT "VALUE OF T, PLEASE"; T

30   INPUT "VALUE OF DELTA T,
     PLEASE"; DELT

40   X1 = FN X(T)

50   X2 = FN X(T + DELT)

60   DELX = X2 - X1

70   XDOT = DELX/DELT

80   PRINT "DERIVATIVE IS"; XDOT

90   END
```

Listing A.2.
Spreadsheet differentiator.

	A	B
1	T	
2	DELT	
3	X(T)	=5*B1^2
4	X(T+DELT)	=5*(B1+B2)^2
5	DELX	=B4-B3
6	XDOT	=B5/B2

grams. I tried to write them as simply as possible to emphasize that differential calculus is in fact simple. Experiment with the programs on your computer. Input various values of Δt while holding t constant to see how it affects the answer that the "digital computer differentiator" provides. Use the programs to experiment with other functions by changing the define function statement in the BASIC program or the statements in cells B3 and B4 in the spreadsheet version.

Even though the programs given in Listings A.1 and A.2 are handy and demonstrate just how easy differential calculus is, I found out early in my career that I saved a lot of time by committing to memory the most frequently used derivatives, provided in Table A.2. You can always look these up in this or other books, but it will take you time and you may not always have your books with you. Every one of these formulas can be derived using equation (A.7), but it's still easier to commit them to memory.

Incidentally, I have been using time t as the *independent variable* and x as the *dependent variable*. That is, $x = f(t)$. I've done this because in many real-life engineering problems, the variables depend on time. However, variables can be a function of another variable that is not time. For example:

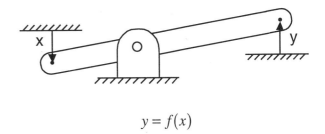

$$y = f(x)$$

Table A.2. Most frequently used derivatives
(where *c* = a constant and
u and *v* are functions of *x*).

$$\frac{dc}{dx} = 0$$

$$\frac{dx}{dx} = 0$$

$$\frac{d}{dx}(u+v) = \frac{du}{dx} + \frac{dv}{dx}$$

$$\frac{d}{dx}(cu) = c\frac{du}{dx}$$

$$\frac{dy}{dx} = \frac{dy}{du} \cdot \frac{du}{dx}$$

$$\frac{d}{dx}\log_a u = \frac{1}{u}\log_a e \frac{du}{dx}$$

$$\frac{d}{dx}\ln u = \frac{1}{u}\frac{du}{dx}$$

$$\frac{d}{dx}(u^n) = nu^{n-1}\frac{du}{dx}$$

$$\frac{d}{dx}(uv) = v\frac{du}{dx} + u\frac{dv}{dx}$$

$$\frac{d}{dx}\left(\frac{u}{v}\right) = \frac{v\frac{du}{dx} + u\frac{dv}{dx}}{v^2}$$

$$\frac{d}{dx}\sin u = \cos u\frac{du}{dx}$$

$$\frac{d}{dx}\cos u = -\sin u\frac{du}{dx}$$

$$\frac{d}{dx}\tan u = \sec^2 u\frac{du}{dx}$$

$$\frac{d}{dx}\cot u = -\csc^2 u\frac{du}{dx}$$

$$\frac{d}{dx}\sec u = \sec u \tan u\frac{du}{dx}$$

$$\frac{d}{dx}\csc u = -\csc u \cot u\frac{du}{dx}$$

$$\frac{d}{dx}a^u = a^u \ln a\frac{du}{dx}$$

$$\frac{d}{dx}e^u = e^u\frac{du}{dx}$$

$$\frac{d}{dx}u^v = vu^{v-1}\frac{du}{dx} + u^v\ln u\frac{dv}{dx}$$

or even two variables that are not functions of time

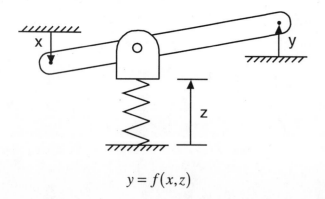

$$y = f(x, z)$$

Nothing changes in the formulas for differentiation except the symbol used for the independent variable. Sometimes you will see a prime symbol (´) used for the shorthand version of differentiation instead of the dot notation. That is,

$$y = f(x)$$

$$\frac{dy}{dx} = y'$$

$$\frac{d^2y}{dx^2} = y''$$

A.3 Integral Calculus

Integral calculus is nothing more than the reverse of differentiation. For example, given that the distance your car is from its starting point is described by $x = 5t^2$, we found that the velocity of the car at any point in time was $v = 0t$ and the acceleration at any point in time was $a = 10$. If integration is the reverse of differentiation, then given the acceleration of the car is 10 ft/sec², we should be able to integrate once and get $v = 10t$ and integrate again and get $x = 5t^2$. Let's look at how we can do this.

We know that the velocity is given by equation (A.8) as

$$v = \frac{dx}{dt} = 10t \qquad \text{(A.8)}$$
repeated

The derivative dx/dt is approximately equal to $\Delta x / \Delta t$ when Δt is very small and was given in equation (A.6) as

$$\frac{\Delta x}{\Delta t} = 10t \qquad \text{as } \Delta t \to 0 \qquad \text{(A.6)}$$
repeated

We can rewrite (A.6) as

$$\Delta x = v\Delta t \qquad \text{(A.17)}$$

Figure A.3 shows the velocity of the car as a function of time and the graphical representation of equation (A.17).

The increment Δx that your car travels in Δt seconds can be seen to be an incremental area under the $v(t)$ curve. If this represents a small part of x, then to get x at any arbitrary time t, all we should have to do is sum all of these small incremental areas up to that time. This can be expressed using the summation symbol Σ as follows:

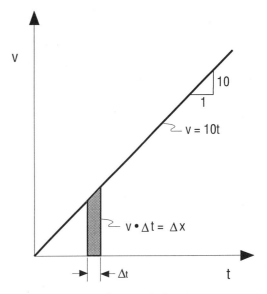

Figure A.3.

$$x \cong \sum_{t=0}^{t=t_1} \Delta x \cong \sum_{t=0}^{t=t_1} v\Delta t \qquad \text{(A.18)}$$

As Δt is made smaller and smaller, the summation symbol (Σ) is replaced by another symbol ($\int$) called the integration symbol and Δx and Δt are replaced with dx and dt. That is,

$$x = \int_{t=0}^{t=t_1} dx = \int_{t=0}^{t=t_1} v\,dt \qquad \text{(A.19)}$$

Since in our example $v = 10t$ we can write

$$x = \int 10t\,dt \qquad \text{(A.20)}$$

We have already seen that one way to evaluate this integral is to find the area under the curve $v(t)$. Another way is to simply find the function whose derivative is $10t$. From our table of derivatives given in Table A.2 we find

$$x = 5t^2 + C \qquad \text{(A.21)}$$

The unknown constant C must be included because when we differentiate equation (A.21) we get

$$\frac{dx}{dt} = 10t \qquad (A.22)$$

no matter what the value of C is. This constant is called *the constant of integration* or *the integration constant*. It must be evaluated from known conditions. In our car example, we said at $t = 0$ that $x = 0$. Thus,

$$x(t = 0) = 5 \cdot 0^2 + C = C \qquad (A.23)$$

So C must equal 0 and we arrive at our answer,

$$x = 5t^2 \qquad (A.24)$$

We can now generalize what we have learned in the following equation

$$\int f(t)dt = F(t) + C \qquad (A.25)$$

In words, given a function $f(t)$, its integral is another function $F(t)$, plus a constant, where $dF(t) / dt = f(t)$. Integration in the form defined by equation (A.25) is called an *indefinite integral* because it does not show the limits over which the integration is to take place. When these limits are shown, we call the integral a *definite integral* and write it as

$$\int_{t=t_1}^{t=t_2} f(t)dt = \left[F(t) + C\right]_{t=t_1}^{t=t_2} = F(t_2) - F(t_1)$$

$$(A.26)$$

The limits over which integration is to take place are defined by t_1 and t_2. The constant of integration drops out.

Sometimes the best way to view integration is with graphs. For example, equation (A.25) indicates that $F(t)$ is equal to $\int f(t)dt$ minus the integration constant. For simplicity, assume the integration constant is zero and look at the following graphs:

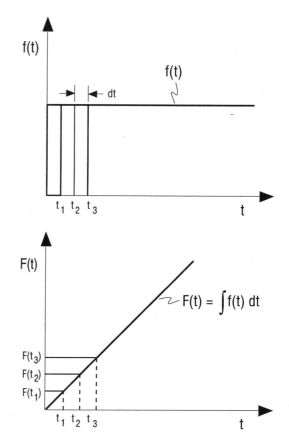

You can see that $F(t)$ is equal to the area under the $f(t)$ curve. Clearly, when t is very small, this area is zero. Each time t increases by an amount dt, the area under the curve increases a constant amount. Since $F(t)$ represents the area under the curve described by $f(t)$ from time $t = 0$ to time t, then $F(t_2) - F(t_1)$ must equal the area under the $f(t)$ curve from $t = 0$ to t_2, minus the area under the $f(t)$ curve from $t = 0$ to t_1. In graphical form:

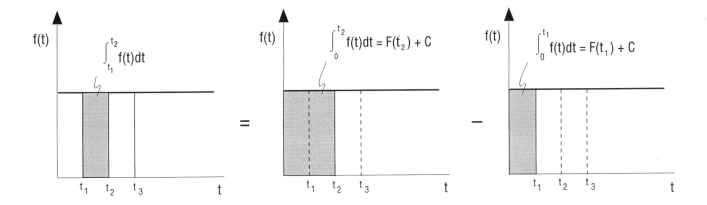

Generally, the independent variable is understood when writing the definite integral and only the limits are shown. That is,

$$\int_{t=t_1}^{t=t_2} f(t)dt = \int_{t_1}^{t_2} f(t)dt = \left[F(t) + C\right]_{t=t_1}^{t=t_2}$$
$$= \left[F(t) + C\right]_{t_1}^{t_2} = F(t_2) - F(t_1)$$

are all equivalent. *Notice that a definite integral is a function of its limits, not a function of the dependent variable t.*

As with differential calculus, integration formulas are available for finding the integral of many functions. I found that I saved a lot of time by committing to memory the formulas given in Table A.3.

You can also double-integrate a function just as you can double-differentiate a function. That is,

$$\iint f(t)dt = F(t) + C_1 t + C_2 \qquad (A.27)$$

Two constants of integration must now be evaluated. For example, in the car example we can take $f(t)$ as

$$f(t) = a = 10 \, ft/\sec^2$$

That is, $f(t)$ is a constant. Double-integrating to get x gives

$$x = \iint 10 \, dt = 5t^2 + C_1 t + C_2 \qquad (A.28)$$

We know that at $t = 0$, $x = 0$ and $v = 0$. Therefore

$$x = 5 \cdot 0^2 + C_1 \cdot 0 + C_2 = 0 \qquad \Rightarrow C_2 = 0 \tag{A.29}$$

and

$$v = 10t_1 + C_1 = 10 \cdot 0 + C_1 = 0 \Rightarrow C_1 = 0 \tag{A.30}$$

We arrive at

$$x = 5t^2$$

as before.

I introduced you earlier to the differentiation operator $D = d/dt$. The inverse *integration operator* is

$$\frac{1}{D} = \int f(t)dt$$

**Table A.3. Most frequently used integrals
(where *c* and *a* are constants and
u and *v* are functions of *x*).**

$$\int du = u + c$$

$$\int u^{-1} du = \int \frac{du}{u} = \ln|u| + c$$

$$\int \sec u \tan u \, du = \sec u + c$$

$$\int (du + dv) = \int du + \int dv$$
$$= u + v + c$$

$$\int \sin u \, du = -\cos u + c$$

$$\int \csc u \cot u \, du = -\csc u + c$$

$$\int \cos u \, du = \sin u + c$$

$$\int a^u du = \frac{a^u}{\ln a} + c$$

$$\int a du = a \int du = au + c$$

$$\int \sec^2 u \, du = \tan u + c$$

$$\int e^u du = e^u + c$$

$$\int u^n du = \frac{u^{n+1}}{n+1} + c$$

$$\int \csc^2 u \, du = -\cot u + c$$

and in block form

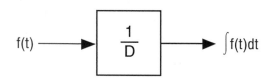

Example:

An "integrator" BASIC computer program is given in Listing A.3 and a spreadsheet version is given in Listing A.4. Carefully review these programs, and experiment with both of them. Change the number of steps required to compute the integral. Notice that you get a different answer each time. Remember that both of these programs give approximations to the integral and are based on equation (A.18). As Δt gets smaller, the answer you get will better approximate the correct value.

Listing A.3. BASIC integrator.

```
5     REM BASIC INTEGRATOR

10    DEF FN X(T) = 10*T^2

20    INPUT "VALUE OF UPPER LIMIT, T2,
      PLEASE";T2

30    INPUT "VALUE OF LOWER LIMIT,
      T1, PLEASE";T1

40    INPUT "NUMBER OF INTEGRATION
      STEPS, PLEASE";N

50    DELT = (T2-T1)/N

60    SUM = 0: T = T1

70    FOR I = 1 TO N

80    SUM = SUM + DELT*FN X(T)

90    T = T + DELT

100   NEXT

110   PRINT "VALUE OF INTEGRAL
      IS";SUM

120   END
```

A.4 Partial Derivatives

The equation for the volume V of a cylinder shows that it is a function of two variables, r and h. That is,

$$V = V(r,h) = \pi r^2 h$$

As I indicated earlier, you can hold one independent variable constant in this equation and then investigate the effect the other has on the dependent variable. You can do this regardless of how many independent variables there are in an equation. In essence, you convert a multiple independent variable function into a single variable function. You can then take derivatives of this function just as you would for a function that had only one independent variable. Such derivatives are called *partial derivatives* and the symbol $\partial u / \partial x$ is used to denote the partial derivative of u with respect to x.

For the general function $u = f(x,y)$, the first partial derivatives are defined as

Listing A.4. Spreadsheet integrator.

	A	B	C	D
1	T1			
2	T2			
3	N			
4	DELT	=(B2-B1)/B3		
5				
6	T	X(T)	X(T)*DELT	SUM
7	=B1	=10*A(7)^2	=B7*$B$4	=0
8	=A7+$B$4	=10*A(8)^2	=B8*$B$4	=D7+C8

$$\frac{\partial u}{\partial x} = \lim_{\Delta x \to 0} \frac{f(x + \Delta x, y) - f(x, y)}{\Delta x} \qquad (A.31)$$

and

$$\frac{\partial u}{\partial y} = \lim_{\Delta y \to 0} \frac{f(x, y + \Delta y) - f(x, y)}{\Delta y} \qquad (A.32)$$

You can see that these definitions are essentially identical to the definition previously given for a function of a single variable. Because the function has two independent variables, there are now two partial derivatives.

Let's take the partial derivatives of the equation for the volume of a cylinder. First, the partial derivative of V with respect to r is:

$$\frac{\partial V}{\partial r} = \lim_{\Delta r \to 0} \frac{V(r + \Delta r, h) - V(r, h)}{\Delta r}$$

$$= \lim_{\Delta r \to 0} \frac{\pi(r + \Delta r)^2 h - \pi r^2 h}{\Delta r}$$

$$= \lim_{\Delta r \to 0} \frac{\pi r^2 h + 2\pi r \Delta r h + \pi(\Delta r)^2 h - \pi r^2 h}{\Delta r}$$

$$= \lim_{\Delta r \to 0} 2\pi r h + \pi \Delta r h$$

$$= 2\pi r h \qquad (A.33)$$

Next take the partial derivative of V with respect to h:

$$\frac{\partial V}{\partial h} = \lim_{\Delta h \to 0} \frac{V(r, h + \Delta h) - V(r, h)}{\Delta h}$$

$$= \lim_{\Delta h \to 0} \frac{\pi r^2 (h + \Delta h) - \pi r^2 h}{\Delta h}$$

$$= \lim_{\Delta h \to 0} \frac{\pi r^2 h + \pi r^2 \Delta h - \pi r^2 h}{\Delta h}$$

$$= \lim_{\Delta h \to 0} \frac{\pi r^2 \Delta h}{\Delta h}$$

$$= \pi r^2 \qquad (A.34)$$

You can see that taking a partial derivative simply involves treating one variable as if it were a constant. You don't have to use the definition equations to compute the partial derivatives; simply recall or refer to the differentiation formulas for a single variable function given in Table A.2.

You can also take higher partial derivatives of multivariable functions. The partial derivatives are written as:

$$\frac{\partial}{\partial x}\left(\frac{\partial u}{\partial x}\right) = \frac{\partial^2 u}{\partial x^2} \qquad (A.35)$$

$$\frac{\partial}{\partial y}\left(\frac{\partial u}{\partial x}\right) = \frac{\partial^2 u}{\partial x \partial y} = \frac{\partial}{\partial x}\left(\frac{\partial u}{\partial y}\right) \qquad (A.36)$$

$$\frac{\partial}{\partial y}\left(\frac{\partial u}{\partial y}\right) = \frac{\partial^2 u}{\partial y^2} \qquad (A.37)$$

Let's take these higher partial derivatives for the volume of a cylinder:

$$\frac{\partial}{\partial r}\left(\frac{\partial V}{\partial r}\right) = \frac{\partial^2 V}{\partial r^2} = 2\pi h \qquad (A.38)$$

$$\frac{\partial}{\partial h}\left(\frac{\partial V}{\partial r}\right) = \frac{\partial^2 V}{\partial h \partial r} = \frac{\partial}{\partial r}\left(\frac{\partial V}{\partial h}\right) = 2\pi r \quad (A.39)$$

$$\frac{\partial}{\partial h}\left(\frac{\partial V}{\partial h}\right) = \frac{\partial^2 V}{\partial h^2} = 0 \qquad (A.40)$$

Increments, Differentials and Total Derivatives

We previously defined the incremental change Δy in a function $y = f(x)$ of a single independent variable x as

$$\Delta y = f(x + \Delta x) - f(x)$$

When Δx is small, the increment Δy is essentially the same as the differential df. So for small values of Δx, we can write

$$\Delta y \cong dy = \frac{df}{dx}dx = \frac{df}{dx}\Delta x \qquad (A.41)$$

For a function $u(x,y)$ of two independent variables the increment Δu is

$$\Delta u = f(x + \Delta x, y + \Delta y) - f(x, y)$$

$$= \left[f(x + \Delta x, y + \Delta y) - f(x, y + \Delta y) \right]$$
$$+ \left[f(x, y + \Delta y) - f(x, y) \right]$$

$$= \frac{\partial u}{\partial x}\Delta x + \frac{\partial u}{\partial y}\Delta y \qquad (A.42)$$

As before, we can replace Δx with dx and Δy with dy and write the total differential as

$$du = \frac{\partial u}{\partial x}dx + \frac{\partial u}{\partial y}dy \qquad (A.43)$$

Let's now apply these equations to the volume of a cylinder. We can write

$$\Delta V = \frac{\partial V}{\partial r}\Delta r + \frac{\partial V}{\partial h}\Delta h \qquad (A.44)$$

Substituting the partial derivatives from above gives

$$\Delta V = (2\pi r h)\Delta r + \left(\pi r^2\right)\Delta h \qquad (A.45)$$

The application of this latter equation should now be clear. If "operating point" values for r and h, say r_o and h_o, are chosen, then this last equation provides the incremental change in volume of the cylinder as a function of the incremental changes in the radius and the height about the operating point r_o, h_o That is,

$$\Delta V = \left(2\pi r_o h_o\right)\Delta r + \left(\pi r_o^2\right)\Delta h \qquad (A.46)$$

You will note that this latter equation is linear in Δr and Δh.

A.5 Taylor's Theorem

You may recall from your algebra that any continuous function $y = f(x)$ can be expanded into an infinite series. We can restrict our attention to a point, $x = a$, and expand the function $f(x)$ about this point in the form

$$f(x) = b_o + b_1(x - a) + b_2(x - a)^2$$
$$+ \ldots + b_n(x - a)^n \qquad (A.47)$$

The coefficients for this equation can be found by taking successive derivatives and then evaluating the derivatives at $x = a$. That is,

$$\frac{df(x)}{dx} = b_1 + 2b_2(x-a) + \ldots + nb_n(x-a)^{n-1}$$
$$\text{(A.48)}$$

$$\frac{d^2f(x)}{dx^2} = 2b_2 + \ldots + n(n-1)b_n(x-a)^{n-2}$$
$$\text{(A.49)}$$

and so forth. Evaluating these at $x = a$ gives

$$f(a) = b_o \qquad \text{(A.50)}$$

$$\left.\frac{df(x)}{dx}\right|_{x=a} = b_1 \qquad \text{(A.51)}$$

$$\left.\frac{d^2f(x)}{dx^2}\right|_{x=a} = 2b_2 \qquad \text{(A.52)}$$

and so forth. Substituting these values back into the expression for $f(x)$ gives

$$f(x) = f(a) + \left.\frac{df(x)}{dx}\right|_{x=a} \times (x-a) + \frac{1}{2}\left.\frac{d^2f(x)}{dx^2}\right|_{x=a}$$
$$\times (x-a)^2 + \ldots + \frac{1}{n!}\left.\frac{d^nf(x)}{dx^n}\right|_{x=a} \times (x-a)^n$$
$$\text{(A.53)}$$

This equation is known as Taylor's Series and it can be proven that the series converges.

One of the most important applications of this equation is associated with the linearization of functions. If we let $\Delta x = x - a$, and use only the first two terms of the Taylor's Series, then any function can be approximated by

$$f(x) = f(a) + \left.\frac{df(x)}{dx}\right|_{x=a} \times \Delta x \qquad \text{(A.54)}$$

Taylor's Series can also be used with multivariable functions. The function $u(x,y)$ can expand about a point (x_o, y_o). Then a linear approximation of the function would be

$$u(x,y) = u(x_o, y_o) + \left.\frac{\partial u(x,y)}{\partial x}\right|_{\substack{x=x_o \\ y=y_o}}$$
$$\times \Delta x + \left.\frac{\partial u(x,y)}{\partial x}\right|_{\substack{x=x_o \\ y=y_o}} \times \Delta y \qquad \text{(A.55)}$$

You can see that this is equivalent to the total derivative given by equation (A.44) earlier.

Appendix
B

The Physics of Work, Power, and Energy in Engineering Systems

B.1 Concepts of Work, Power, and Energy in Electrical Elements

Voltage is defined as the work that must be done to move a unit of electrical charge from one point to another. Therefore we can write voltage points 1 and 2 as

$$V_{21} = \frac{dW_{21}}{dq} \qquad \text{(B.1)}$$

where dW_{21} is the work that must be done to move the electrical charge dq from point 1 to point 2. A unit of measure for work is the *joule*. Equation (B.1) then defines volts as joules per coulomb. That is, one volt is equal to one joule of work per one coulomb of charge. One joule of work is equivalent to one watt-sec or 0.737 ft-lbs.

Current is defined as the flow of electrical charge per unit of time. This can be written in the form of a derivative. That is,

$$i = \frac{dq}{dt} \qquad \text{(B.2)}$$

The units which apply to this equation are coulombs per second, or amperes.

Power is defined as the rate at which work is performed. This too can be written in the form of a derivative. That is,

$$P = \frac{dW}{dt} \qquad \text{(B.3)}$$

The product of voltage differential across an electrical element and the current flowing through the element is equal to power. This is a very important concept. You can see this by multiplying equation (B.1) and (B.2). That is,

$$V_{21} \times i = \frac{dW_{21}}{dq} \times \frac{dq}{dt} = \frac{dW_{21}}{dt} = P \qquad \text{(B.4)}$$

The units for power are

$$P = V_{21}i = 1 \text{ volt} \times 1 \text{ ampere}$$

or

$$P = V_{21}i = 1\frac{\text{joule}}{\text{coulomb}} \times 1\frac{\text{coulomb}}{\text{second}}$$

$$= \frac{\text{joules}}{\text{second}} = 1 \text{ watt}$$

Since work is a transitory form of energy, we can rewrite equation (B.3) as

$$\frac{dE}{dt} = P \qquad \text{(B.5)}$$

Equation (B.5) can be integrated to obtain the energy stored in, or dissipated by, an electrical element over a time interval from $t = t_a$ to $t = t_b$. That is,

$$
\begin{aligned}
dE &= Pdt \\
E &= \int Pdt \\
E_b - E_a &= \int_{t_a}^{t_b} Pdt = \int_{t_a}^{t_b} (V_{21}i)dt \qquad \text{(B.6)}
\end{aligned}
$$

B.2 Concepts of Work, Power, and Energy in Translational Mechanical Elements

If a force F is applied over an incremental distance ds in the direction of the force, an amount of work dW is done equal to

$$dW = Fds \qquad \text{(B.7)}$$

The units for this equation are force times distance. If the force is measured in pounds and the distance in feet, then the units of work are ft-lbs.

Equation (B.7) can also be written in terms of velocity of a point moving in the direction of the force. That is,

$$dW = F\frac{ds}{dt}dt = Fvdt \qquad (B.8)$$

The amount of work done in an interval from $t = t_a$ to $t = t_b$ can be obtained by integrating equation (B.8). That is,

$$W = \int_{t_a}^{t_b} Fvdt \qquad (B.9)$$

Since power is the rate at which work is performed, it is also clear from equation (B.8) that

$$Power = \frac{dW}{dt} = Fv \qquad (B.10)$$

B.3 Concepts of Work, Power, and Energy in Rotational Mechanical Elements

The concepts of work, power, and energy for rotational mechanical elements are the same as those for translational mechanical components, except that torque and angular velocity are used in lieu of force and linear velocity. We can write the power as

$$Power = Q\omega_{21} \qquad (B.11)$$

and work done between $t = t_a$ and $t = t_b$ as

$$W_{ab} = \int_{t_a}^{t_b} Q\omega_{21}dt \qquad (B.12)$$

B.4. Concepts of Work, Power, and Energy in Fluid Elements

Refer to the stream tube shown in Figure B.1.

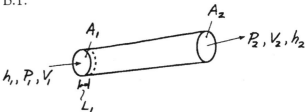

Figure B.1.

The work that must be done to force the fluid into the entrance of the tube at point 1 is equal to the product of the force p_1A_1 and the distance l_1. That is

$$W_1 = p_1A_1l_1 \qquad (B.13)$$

Let A_1l_1 equal the volume vol_1 and rewrite this equation as

$$W_1 = p_1vol_1 \qquad (B.14)$$

In a similar manner, the work done by the fluid exiting the streamtube can be written as

$$W_2 = p_2vol_2 \qquad (B.15)$$

The net work done is then

$$W_1 - W_2 = p_1vol_1 - p_2vol_2 \qquad (B.16)$$

Recalling that power is the rate at which work is done, we can express equation (B.14) in terms of the volume rate of flow. That is,

$$Power_1 = \frac{dW_1}{dt} = p_1\frac{dvol}{dt} = p_1q_v \qquad (B.17)$$

Similarly,

$$Power_2 = p_2 q_v \qquad \text{(B.18)}$$

Subtracting $Power_2$ from $Power_1$ gives the net power as the product of the pressure difference across the tube ends and the volume rate of flow through the tube:

$$Power_1 - Power_2 = p_1 q_v - p_2 q_v = p_{12} q_v$$
$$\text{(B.19)}$$

As we did with the electrical and mechanical elements, we can obtain the energy stored in, or dissipated by, a fluid element over a time interval from $t = t_a$ to $t = t_b$. That is,

$$dE = (Power)dt$$

$$E = \int (Power)dt$$

$$E_b - E_a = \int_{t_a}^{t_b} (Power)dt = \int_{t_a}^{t_b} (p_{12} q_v)dt \qquad \text{(B.20)}$$

B.5. Concepts of Work, Power and Energy in Thermal Elements

When mechanical work is done on a body, its temperature will rise unless heat is removed from the body. The First Law of Thermody-namics states that work and heat can be converted from one to the other. Consequently, the units of heat and work are equivalent. In the English system, temperature is usually measured in Fahrenheit degrees and heat in British Thermal Units (BTU). One BTU is equivalent to 778.172 ft-lbs. In the SI system, temperature is measured in degrees Centigrade and heat in joules. (One joule is equal to one watt-sec or 0.737 ft-lbs.)

There are three basic ways in which heat can be transferred from a hotter to a colder body: *convection*, *radiation* and *conduction*. Convection-type heat transfer takes place primarily in liquids and gases. Heat is transferred as the result of matter moving from one location to another due to currents set up by the temperature differentials. Radiation-type heat transfer takes place as a result of energy carried by electromagnetic waves. Conduction-type heat transfer usually involves substances in the solid phase. Heat is transferred at the atomic level without any visible motion of matter.

Unlike the other types of systems being considered, in thermal systems, power is *not* equal to the product of the across variable (temperature) and the through variable (heat flow rate). The through variable itself—heat flow rate—is power.

Index